DESIGN OF JOINTS IN STEEL STRUCTURES

UK EDITION

ECCS EUROCODE DESIGN MANUALS

ECCS EDITORIAL BOARD
Luís Simões da Silva (ECCS)
António Lamas (Portugal)
Jean-Pierre Jaspart (Belgium)
Reidar Bjorhovde (USA)
Ulrike Kuhlmann (Germany)

DESIGN OF STEEL STRUCTURES – 2ND EDITION
Luís Simões da Silva, Rui Simões and Helena Gervásio

FIRE DESIGN OF STEEL STRUCTURES – 2ND EDITION
Jean-Marc Franssen and Paulo Vila Real

DESIGN OF PLATED STRUCTURES
Darko Beg, Ulrike Kuhlmann, Laurence Davaine and Benjamin Braun

FATIGUE DESIGN OF STEEL AND COMPOSITE STRUCTURES
Alain Nussbaumer, Luís Borges and Laurence Davaine

DESIGN OF COLD-FORMED STEEL STRUCTURES
Dan Dubina, Viorel Ungureanu and Raffaele Landolfo

DESIGN OF JOINTS IN STEEL AND COMPOSITE STRUCTURES
Jean-Pierre Jaspart and Klaus Weynand

DESIGN OF STEEL STRUCTURES FOR BUILDINGS IN SEISMIC AREAS
Raffaele Landolfo, Federico Mazzolani, Dan Dubina, Luís Simões da Silva and Mario d'Aniello

ECCS – SCI EUROCODE DESIGN MANUALS

DESIGN OF JOINTS IN STEEL STRUCTURES, UK EDITION
Jean-Pierre Jaspart and Klaus Weynand
Adapted to UK by Graham Couchman and Ana M. Girão Coelho

DESIGN OF STEEL STRUCTURES, UK EDITION
Luís Simões da Silva, Rui Simões, Helena Gervásio
Adapted to UK by Graham Couchman

ECCS EUROCODE DESIGN MANUALS – BRAZILIAN EDITIONS

DIMENSIONAMENTO DE ESTRUTURAS DE AÇO
Luís Simões da Silva, Rui Simões, Helena Gervásio, Pedro Vellasco, Luciano Lima

INFORMATION AND ORDERING DETAILS

For price, availability, and ordering visit our website **www.steelconstruct.com**.
For more information about books and journals visit **www.ernst-und-sohn.de**.

DESIGN OF JOINTS IN STEEL STRUCTURES

**Eurocode 3: Design of steel structures
Part 1-8 – Design of Joints**

UK Edition

**Jean-Pierre Jaspart
Klaus Weynand**

Adapted for the UK by

**Graham Couchman
Ana M. Girão Coelho**

Design of Joints in Steel Structures

UK Edition
2017

Published by:
ECCS – European Convention for Constructional Steelwork
publications@steelconstruct.com
www.steelconstruct.com

Sales:
Wilhelm Ernst & Sohn Verlag für Architektur und technische Wissenschaften
GmbH & Co. KG, Berlin

ISBN (ECCS): 978-92-9147-135-5
ISBN (Ernst & Sohn): 978-3-433-03216-9
Legal dep.: 423588/17 Printed in Sersilito, Empresa Gráfica Lda, Maia, Portugal
Photo cover credits: Image courtesy of Mark Lawson, SCI

TABLE OF CONTENTS

FOREWORD	xiii
PREFACE	xv
UK FOREWORD	xix
NOTATION	xxix

Chapter 1

INTRODUCTION	1
1.1 General	1
1.1.1 Aims of the book	1
1.1.2 Brief description of the contents of the book	10
1.1.3 Types of structural systems and joints covered	11
1.1.4 Basis of design	12
1.2 Definitions	12
1.2.1 Joint properties	14
1.2.2 Sources of joint deformation	15
1.2.3 Beam splices and column splices	19
1.2.4 Beam-to-beam joints	20
1.2.5 Column bases	21
1.2.6 Hollow section joints	22
1.3 Material choice	22
1.4 Fabrication and erection	26
1.5 Costs	27
1.6 Application of the "static approach"	27
1.6.1 Component approach	29
1.6.2 Hybrid joints aspects	35
1.7 Design tools	35
1.7.1 Types of design tools	36

v

1.7.2   Examples of design tools   37

1.8   Worked examples   40

Chapter 2

STRUCTURAL ANALYSIS AND DESIGN   43

2.1   Introduction   43
       2.1.1   Elastic or elastoplastic analysis and verification process   44
       2.1.2   First order or second order analysis   45
       2.1.3   Integration of joint response into the frame analysis and
       design process   46

2.2   Joint modelling   47
       2.2.1   General   47
       2.2.2   Modelling and sources of joint deformation   49
       2.2.3   Simplified modelling according to EN 1993   50
       2.2.4   Concentration of the joint deformation   51

2.3   Joint idealisation   56
       2.3.1   Elastic idealisation for an elastic analysis   57
       2.3.2   Rigid-plastic idealisation for a rigid-plastic analysis   58
       2.3.3   Non-linear idealisation for an elastic-plastic analysis   59

2.4   Joint classification   59
       2.4.1   General   59
       2.4.2   Classification based on mechanical joint properties   60

2.5   Ductility classes   62
       2.5.1   General concept   62
       2.5.2   Requirements for classes of joints   65

Chapter 3

CONNECTIONS WITH MECHANICAL FASTENERS   67

3.1   Mechanical fasteners   67

3.2   Categories of connections   69
       3.2.1   Shear connections   69
       3.2.2   Tension connections   71

3.3	Positioning of bolt holes	72
3.4	Design of the basic components	74
	3.4.1 Bolts in shear	74
	3.4.2 Bolts in tension	75
	3.4.3 Bolts in shear and tension	76
	3.4.4 Preloaded bolts	77
	3.4.5 Plates in bearing	85
	3.4.6 Block tearing	86
	3.4.7 Injection bolts	87
	3.4.8 Pins	88
	3.4.9 Blind bolting	91
	3.4.10 Nails	94
	3.4.11 Eccentricity of angles	95
3.5	Design of connections	97
	3.5.1 Bolted lap joints	97
	3.5.2 Bolted T-stubs	101
	3.5.3 Gusset plates	113
	3.5.4 Long joints	117

Chapter 4

WELDED CONNECTIONS 119

4.1	Types of welds	119
	4.1.1 Butt welds	119
	4.1.2 Fillet welds	120
	4.1.3 Fillet welds all round	121
	4.1.4 Plug welds	122
4.2	Construction constraints	122
	4.2.1 Mechanical properties of materials	122
	4.2.2 Welding processes, preparation of welds and weld quality	123
	4.2.3 Geometry and dimensions of welds	127
4.3	Design of welds	130
	4.3.1 Generalities	130
	4.3.2 Fillet welds	131
	4.3.3 Fillet welds all round	134

		4.3.4	Butt welds	135
		4.3.5	Plug welds	136
		4.3.6	Concept of full strength fillet weld	136
	4.4	Distribution of forces in a welded joint		139
		4.4.1	Generalities	139
		4.4.2	Particular situations	141

Chapter 5

SIMPLE JOINTS   **147**

5.1	Introduction		147
5.2	Beam-to-column and beam-to-beam joints		149
	5.2.1	Introduction	149
	5.2.2	Scope and field of application	150
	5.2.3	Joint modelling for frame analysis and design requirements	153
	5.2.4	Design resistance	156
	5.2.5	Practical ways to satisfy the ductility and rotation requirements	163
5.3	Column bases		174
	5.3.1	Introduction	174
	5.3.2	Basis for the evaluation of the design resistance	176
	5.3.3	Resistance to axial forces	177
	5.3.4	Resistance to shear forces	185

Chapter 6

MOMENT-RESISTING JOINTS   **189**

6.1	Introduction		189
6.2	Component characterisation		190
	6.2.1	Column web panel in shear in steel or composite joints	190
	6.2.2	Column web in transverse compression in steel or composite joints	192
	6.2.3	Column web in transverse tension	196
	6.2.4	Column flange in transverse bending	197
	6.2.5	End plate in bending	203

6.2.6   Flange cleat in bending   205

6.2.7   Beam or column flange and web in compression   207

6.2.8   Beam web in tension   209

6.2.9   Plate in tension or compression   210

6.2.10 Bolts in tension   211

6.2.11 Bolts in shear   212

6.2.12 Bolts in bearing (on beam flange, column flange, end plate or cleat)   213

6.2.13 Concrete in compression including grout   213

6.2.14 Base plate in bending under compression   214

6.2.15 Base plate in bending under tension   214

6.2.16 Anchor bolts in tension   215

6.2.17 Anchor bolts in shear   215

6.2.18 Anchor bolts in bearing   215

6.2.19 Welds   216

6.2.20 Haunched beam   216

6.3   Assembly for resistance   217

6.3.1   Joints under bending moments   217

6.3.2   Joints under axial forces   225

6.3.3   Joints under bending moments and axial forces   226

6.3.4 $M - N - V$ 233

6.3.5   Design of welds   234

6.4   Assembly for rotational stiffness   238

6.4.1   Joints under bending moments   238

6.4.2   Joints under bending moments and axial forces   247

6.5   Assembly for ductility   249

6.5.1   Steel bolted joints   250

6.5.2   Steel welded joints   252

6.6   Application to steel beam-to-column joint configurations   253

6.6.1   Extended scope   253

6.6.2   Possible design simplifications for end plate connections   256

6.6.3   Worked example   257

6.7   Application to steel column splices   281

6.7.1   Common splice configurations   281

ix

6.7.2   Design considerations   283

6.8   Application to column bases   284
      6.8.1   Common column basis configurations   284
      6.8.2   Design considerations   287

Chapter 7

LATTICE GIRDER JOINTS   295

7.1   General   295

7.2   Scope and field of application   296

7.3   Design models   298
      7.3.1   General   298
      7.3.2   Failure modes   299
      7.3.3   Models for CHS chords   300
      7.3.4   Model for RHS chords   301
      7.3.5   Punching shear failure   302
      7.3.6   Model for brace failure   303
      7.3.7   *M-N* interaction   304

Chapter 8

JOINTS UNDER VARIOUS LOADING SITUATIONS   305

8.1   Introduction   305

8.2   Joints in fire   306

8.3   Joints under cyclic loading   307

8.4   Joints under exceptional events   308

Chapter 9

DESIGN STRATEGIES   311

9.1   Introduction   311

9.2   Traditional design approach   314

9.3   Integrated design approach   317

9.4   Economic considerations   319

9.4.1   Fabrication and erections costs   319

9.4.2   Savings of material costs   322

9.4.3   Summary and conclusions   323

BIBLIOGRAPHIC REFERENCES   325

Annex A   Practical values for required rotation capacity   337

Annex B   Values for lateral torsional buckling strength of a fin plate   339

FOREWORD

With the ECCS book "Joints in Steel and Composite Structures" the authors succeeded in placing the joints on the rightful place they deserve in the structural behaviour of steel and composite steel-concrete structures. The many times used word "details" for the joints in structures by far underestimates the importance of joints in the structural behaviour of buildings and civil engineering structures. In their chapter "Aim of the book" the authors clearly explain how the design and safety verification of structures runs in an integral manner where all structural components, including the joints, play balanced roles leading to economic structures.

This book can be seen as a background document for Eurocode 3 "Design of Steel Structures" as far as it concerns structural joints. The central theme in describing the behaviour of joints is using the component method and this is leading all over in this book. The book contain many aspects such as design, fabrication, erection and costs.

In this book attention is paid on joint modelling and idealisation, joint classification for strength and stiffness and deformation capacity. This all for connections with mechanical fasteners and for welded connections, for simple joints and moment resistant joints. Also lattice girder joints are described.

The book provides the designer with design strategies to arrive at economic structures.

The authors based themselves on many bibliographic references covering a time span of about 65 years. Many of these references present research of the authors themselves and of the other members of the ECCS-Technical Committee TC10 "Structural Connections".

It was really a privilege to have been the chairperson of this committee from 1998 till the end of 2012 and I thank the authors Prof. Dr. Ir. Jean-Pierre Jaspart and Dr.-Ing. Klaus Weynand for their large effort in writing this book.

Prof. ir. Frans Bijlaard

PREFACE

Steel constructions and composite steel-concrete constructions are generally erected on site by the assembly of prefabricated structural parts prepared in the workshop. These parts may themselves be the result of an assembly of individual elements. An example is the assembly by bolting on site of fabricated sections welded in the workshop.

In these construction types, joints and connections play a key role and recommendations and guidelines are required for engineers and constructors faced with the conception and design, the fabrication and the erection of such structures. In the Structural Eurocodes, these aspects are mainly covered in the execution standard EN 1090-2 and in the design standard EN 1993-1-8 (Eurocode 3, Design of steel structures, Part 1-8: Design of Joints).

In the present book which is part of the series of ECCS Eurocode Design Manuals, the main focus is given to design aspects, but references are also made to EN 1090-2 when necessary.

In comparison to other fields, the design procedures for joints and connections have significantly evolved in the last decades as a result of the progressive awareness by practitioners of the significant contribution of joints and connections to the global cost of structures. Design for low fabrication and erection costs as well as high resistance is therefore the targeted objective of modern design codes, the achievement of which has justified the development of new calculation approaches now integrated into the two afore-mentioned Eurocodes. This situation justifies the writing of the present manual with the main goal to demystify the process by explaining the new concepts for the design of joints and to integrate their mechanical response into the structural frame analysis and design process, by providing "keys" for a proper application in practice and finally by providing well documented worked examples.

To refer to "modern" or "new" design approaches and philosophies does not mean that traditional ways are out dated or no more valid. It should be understood that the design methods recommended in the Eurocodes are a collection of European practices including the results of intensive research efforts carried out in the last decades and so give many options and alternatives to engineers who are trying to develop safe and economic solutions.

Chapter 1 introduces generalities about joint properties, aspects of materials, fabrication, erection and costs, design approach - and especially the so-called component method - and design tools available to practitioners for easier code application. The integration of the response of the joints into the structural analysis and design process is addressed in chapter 2. In chapter 3, attention is paid to the design of common connections with mechanical fasteners. Preloaded bolts and non-preloaded bolts are mainly considered but the use of some less classical connectors is also briefly described. Welded connections are covered in chapter 5.

The next three chapters relate to three specific types of joints, respectively simple joints, moment resisting joints and lattice girder joints. For these, substantial innovations are brought in the Eurocodes in comparison to traditional national codes; particularly for simple and moment resisting joints. A significant number of pages is therefore devoted to these topics in this manual.

The design of joints under static loading, as it is addressed in the seven first chapters, is essential in all cases but further checks or different conceptual design of the joints are often required in case of load reversal, fire, earthquake or even exceptional events like impact or explosion. Chapter 8 summarises present knowledge in this field.

Traditionally joints were designed as rigid or pinned, which enables the design of the frame, and the design of the joints to be treated independently. The clear economical advantage associated in many situations with the use of semi-rigid and/or partial-strength joints leads however to "structure-joints" interactions that have to be mastered by the engineer so as to fully profit from the potential cost savings. The Eurocodes do not cover this interaction. From this point of view, chapter 9 may be considered as "a première" even if the content has already been partly described years ago in an ECSC publication.

Before letting the reader discover the contents of this book, we would like to express acknowledgment. We are very grateful to Prof. Frans Bijlaard for all the comments, suggestions and corrections he made through the review process of the present manual. Warm thanks are also addressed to José Fuchs and Sönke Müller who helped us in preparing the drawings. Last but not least we would like to thank our wives for their patience when we worked "on our project" during innumerable evenings and weekends.

Jean-Pierre Jaspart
Klaus Weynand

UK FOREWORD

INTRODUCTION

This book is the second in a series of joint publications of the European Convention for Constructional Steelwork (ECCS) and The Steel Construction Institute (SCI). It will help provide UK engineers and designers with a better understanding of the design principles in the European Code of Practice for the Design of Steel Structures (EN 1993 – Eurocode 3), and the individual Part 1.8 (EN 1993-1-8) for the design of steel joints, in particular. Reference will also be made to the associated Eurocodes for concrete (EN 1992 – Eurocode 2), and earthquake engineering (EN 1998 – Eurocode 8). The book complements other SCI design guides, e.g. SCI (2011 and 2013a).

The current publication is a first UK edition based on the first edition of the ECCS manual for the *Design of Joints in Steel and Composite Structures*, which was published in 2016. The scope of this UK edition is narrower than the original publication in two specific areas that are felt to be of limited interest to current practice in the UK; there is no content covering composite joints, and chapter 9 (covering frame design strategies) has been shortened to simply give an overview of strategies. In this edition, the reader will find information that is either of a general nature, or relevant to specific sections of the publication, to facilitate its application in a UK context.

The content includes useful background to the code rules, and a reminder of some engineering principles. Information is presented in the context of the Eurocodes terminology and notation, and with reference to clause numbers, etc. to help the reader's familiarity with EN 1993-1-8. The document that serves as the basis for content and numbering of clauses is the UK implementation of EN 1993-1-8:2005 incorporating corrigenda December 2005 and July 2009. At the time of writing (Autumn 2016), part 1-8 of EN 1993 is being revised with publication anticipated after 2020.

Finally it is noted and should be accepted that there will inevitably be some differences of interpretation between the ECCS recommendations and those previously published by the SCI.

GENERAL COMMENTS

Design of joints (connections) in steel structures in the UK is covered by BS EN 1993-1-8 and its national Annex (NA) (BSI, 2010). BS EN 1993-1-8 has clear definitions for connections and joints. The term *joint* refers to the zone where members are interconnected, and a connection is the location where elements meet. The distinction in terminology is usually emphasised in Continental Europe but traditional practice in the UK is to use the generic term *connection*.

The following partial safety factors are defined in the UK NA.

Partial factor	Value	Comment
γ_{M2}	1.25	Used for the resistance of bolts, pins, and welds
γ_{M2}	1.25	Used for the resistance of plates in bearing ($\gamma_{M2} = 1.5$ should be used if deformation control is important)
γ_{M3}	1.25	Used for slip resistance at Ultimate Limit States
$\gamma_{M3,serv}$	1.10	Used for slip resistance at Serviceability Limit States (SLS)
γ_{M4}	1.00	Used for the resistance of injection bolts
γ_{M5}	1.00	Used for the resistance of joints in hollow section lattice girders
$\gamma_{M6,serv}$	1.00	Used for the resistance of pins at SLS
γ_{M7}	1.10	Used for the preload of high strength steel bolts conforming to BS EN 14399-4 and BS EN 14399-8 (BSI, 2015 and 2007)
	1.00	Used for the preload of (other) high strength steel bolts

Specific reference to the UK Building Regulations is also made in this UK edition. The UK Building Regulations code has structural integrity requirements in order to avoid disproportionate collapse, and ensure robustness of the complete structure. *Disproportionate collapse* can be defined as a region of structural collapse out of proportion to the initiating cause of failure. The most relevant to connection/joint design is that the detailing for every beam member must be adequate to resist a minimum horizontal force at ultimate failure (or *tying resistance*). Guidance on the design values of tying forces can be found in BS EN 1991-1-7 Annex A and its UK annex (BSI, 2014).

SECTION SPECIFIC COMMENTS

Section 1.1.1

UK practice normally considers only two main forms of joint modelling, simple and continuous. Ideally frames designed on the basis of *simple joint modelling* should use joints between members that possess negligible rotational stiffness, and transmit the beam reactions in shear into the columns without developing significant moments. These joints may be treated as perfect pins. Members can then be designed in isolation, either as (predominantly) axially loaded columns, or as simply supported beams. Joints designed within the principles of *continuous modelling* are capable of transmitting significant moments and are able to maintain the original angle between adjacent members virtually unchanged. This form of construction of steel frames is particularly advantageous when beam deflections are critical or if bracing systems are not possible.

To comply with one or the other design assumption, steelwork connections are detailed as *simple* or *moment-resisting* joints. Because joints in frame structures are often treated by adopting a degree of standardisation, the designer usually details the connections in accordance with the principles in the series of SCI Green Books, which present this material in the form of step-by-step design procedures and tables covering standard arrangements.

xxi

Other approximate analysis methods continue to be used among British designers, the most representative being the *wind-moment method* for the design of unbraced multi-storey steel frames. The method assumes that (i) under gravity loads the beam-to-column joints act as pinned connections, and (ii) under horizontal wind loads these joints are rigid, although it lacks transparency. This method can be seen as a manifestation of the semi-continuous principles. Wind moment connection details can be found in SCI (1995a). With the now widespread use of software in design offices for rigorous second-order analysis, it is doubtful if the use of the wind-moment method should be advocated any longer, other than for initial sizing.

Section 1.3

Although S460 is attracting growing interest, S355 is still the dominant grade of steel used in the UK.

Section 1.6.1

With reference to Figure 1.20 it is noted that (site) welded joints and angle flange cleats are rarely, if ever, used in the UK.

Section 1.6.2

Hybrid connections are rarely used in the UK for the reasons noted.

Section 1.7.2

There is not much use of CoP$^©$ in the UK. There are a number of other software tools on the market, some more comprehensive than others.

Section 2.2

Although in UK design practice most aspects related to joint modelling have developed to a high degree of sophistication, proper consideration of the structural interaction between the joints and members is rarely fully taken into account. It is indeed very difficult to predict and model the complex behaviour of a joint (strength, stiffness and rotation capacity). Traditional UK practice is to predict joint behaviour on the basis of past experience.

BSI (2010) separates the modelling of the rotational behaviour of the connection elements (that is, the bolts, the welds, the plates, etc.) and that of the column web panel in shear to characterise the overall joint behaviour. As a simplification, the two sources of joint deformation are combined by means of the so-called transformation parameter β. This coefficient is related to the web panel internal actions and can be calculated from the moments at each side of the web panel. The β parameter is currently only defined for same-depth beams which meet at the same level. In many practical double-sided joints, beams will not be of the same depth, or at the same height. BS EN 1993-1-8 is silent on such joints.

Section 2.3

As explained above, typical UK design practice does not consider this sophisticated procedure for idealisation of joints.

Section 2.4

The NA to BS EN 1993-1-8 gives practical guidance to classify joints. It also states the following:

> *"Until experience is gained with the numerical method of calculating rotational stiffness given in BS EN 1993-1-8:2005, 6.3 and the classification by stiffness method given in BS EN 1993-1-8:2005, 5.2.2, semi-continuous elastic design should only be used where either it is supported by test evidence according to BS EN 1993-1-8:2005, 5.2.2.1(2) or where it is based on satisfactorily performance in a similar situation."*

Evidence on values of column base stiffness for practical global analysis is given in SCI (2006).

Section 2.5

The concept of ductility classes is not widely applied in the UK. It is generally assumed that standard connection details to the SCI Green Books provide enough rotation capacity and so behave in a ductile manner, as evidenced by physical testing and experience.

Section 3.1

The NA to BS EN 1993-1-8 excludes bolts of classes 5.8 and 6.8. It also states that bolt class 4.8 may only be used if the manufacturing process (i) is in accordance with BS EN ISO 898-1 (BSI, 2013), and (ii) includes suitable procedures to avoid hydrogen embrittlement.

Section 3.2.2

The NA to BS EN 1993-1-8 states that if the preload is not explicitly used in the design calculations, then no specific level of preload is required.

Section 3.4.7

Injection bolts are rarely used in the UK.

Section 3.4.8

Reference is made to SCI (1995b) to calculate the bending moments in the pin of a pin bearing.

Section 3.4.9

Flow drill blind bolting is not common practice in the UK. SHS blind bolting is a common system, but there are several manufactures other than that suggested in the book.

Section 3.4.10

Nails are rarely used in the UK.

Section 3.5.2

T-stub configurations with four bolts in a row are rarely used in the UK.

Section 3.5.3

Reference is made to the SCI method for the design of gusset plates, see SCI (2011).

Section 4.2.1

It is anticipated that specific rules given in the current BS EN 1993-1-12 (BSI, 2009) for high strength steels will be incorporated in the revised BS EN 1993-1-8, to appear after 2020.

Section 5.2.2

"Header plates" are known as "partial depth end plates" in the UK. Web cleat connections are no longer commonly used in the UK. The full depth end plate connection shown on the left of Figure 5.9 is classified as a simple joint in the UK (SCI, 2011).

Section 5.2.3

The UK Building Regulations require that joints are designed for tying forces in order to satisfy structural integrity requirements. Guidance on the design values of tying forces is given in BS EN 1991-1-7 Annex A and its UK NA. The requirements relate to the building class, with a design value of horizontal force generally not less than 75 kN, and usually significantly higher. Simple joints detailed to SCI (2011) satisfy these requirements.

Section 5.2.5

It is generally safe to assume that the ductility (rotation requirements) are satisfied provided that joints are detailed in accordance with SCI (2011).

Section 5.3.1

The column base detailing on the left of Figure 5.21 is not commonly used in the UK. We never recommend less than four bolts – they are needed to assure stability and facilitate location during erection. The anchoring systems shown in Figure 5.22 are not usually adopted in the UK. Reference is made to SCI (2013a) for common anchor systems.

Section 6.2

Publication SCI (2013b) summarises the principal differences between BS 5950 and BS EN 1993-1-8 (BSI, 2000 and 2010) in terms of component characterisation.

With reference to Figure 6.6, it is anticipated that the revised BS EN1993-1-8 will include analytical expressions for the definition of the so-called α coefficient.

Rules of thumb for the characterisation of the ductility of some individual components are given in SCI (1995a).

Section 6.3.1

The UK NA to BS EN 1993-1-8 gives specific rules for the full plastic distribution of effective design tension resistance. The procedure is also summarised in SCI (2013a).

Section 6.3.3

Where a joint is subjected to combined bending moments and axial forces, the concept of an *applied moment modified by axial load* can be used, see SCI (1995a).

Section 6.4

Rotational stiffness is not normally evaluated in UK design practice, which usually relies on past experience.

Section 6.5

Some rules of thumb for the characterisation of the ductility of moment-resisting joints are given in SCI (1995a).

Section 6.6.1

With reference to Figure 6.36 it is noted that this joint configuration is hardly (if ever) used in the UK.

Section 6.6.3

It should be noted that European sections are not commonly used in the UK, although there is growing interest in the market. Steel grade S235 is not used in the UK, S355 being the current common grade. Combinations of plates of 15 mm thickness with M24 bolts are common. This combination is assumed to provide enough rotation capacity thus satisfying ductility requirements, because ductile failure of the plates, not brittle failure of the bolts, governs the behaviour.

Section 6.7.2

Standard practice in the UK does not account for this 25% rule. Column splices are usually detailed to SCI (2011) to ensure continuity of stiffness through the splice so that the member design is not invalidated.

Section 8.4

The UK Building Regulations require that joints are designed for tying forces in order to satisfy structural integrity requirements. Guidance on the design values of tying forces is given in BS EN 1991-1-7 Annex A and its UK NA (BSI, 2014).

Section 9.1

In the UK the "fabricators" are known as "Steelwork contractors" because they do more than just fabricate.

REFERENCES TO NATIONAL FOREWORD

BSI (2000). BS 5950-1:2000 – Structural use of steelwork in building – Part 1: Code of Practice for Design: rolled and welded sections, *British Standards Institution*, United Kingdom.

BSI (2007). BS EN 14399-8:2007 – High-strength structural bolting assemblies for preloading. System HV. Hexagon fit bolt and nut assemblies, *British Standards Institution*, United Kingdom.

BSI (2009). BS EN 1993-1-12:2007 – Eurocode 3: Design of steel structures – Part 1.12: Additional rules for the extension of EN 1993 up to steel grades S 700, *British Standards Institution*, United Kingdom.

BSI (2010). BS EN 1993-1-8:2005 – Eurocode 3: Design of steel structures – Part 1.8: Design of joints, *British Standards Institution*, United Kingdom.

BSI (2013). BS EN ISO 898-1:2013 – Mechanical properties of fasteners made of carbon steel and alloy steel. Bolts, screws and studs with specified property classes. Coarse thread and fine pitch thread, *British Standards Institution*, United Kingdom.

BSI (2014). BS EN 1991-1-7:2006+A1:2014 – Eurocode 1. Actions on structures – Part 197: General actions, *British Standards Institution*, United Kingdom.

BSI (2015). BS EN 14399-4:2015 – High-strength structural bolting assemblies for preloading. System HV. Hexagon bolt and nut assemblies, *British Standards Institution*, United Kingdom.

SCI (1995a). Joints in steel construction. Moment connections, P207, *The Steel Construction Institute.*

SCI (1995b). Advisory Desk Notes: Bending Moment in a pin, AD 172, *The Steel Construction Institute.*

SCI (2006). Column base stiffness for global analysis, NCCI-SN045a, *The Steel Construction Institute.*

SCI (2011). Joints in steel construction. Simple joints to Eurocode 3, P358, *The Steel Construction Institute.*

SCI (2013a). Joints in steel construction. Moment-resisting joints to Eurocode 3, P398, *The Steel Construction Institute.*

SCI (2013b). Interim Report: Design of Portal Frames to Eurocode 3: An overview for UK designers, P400, *The Steel Construction Institute.*

NOTATION

LOWER CASES

a	throat thickness of a fillet weld
$a_{eq,R}$	width of a T-stub in compression
b	section breadth (subscript b or c refers to beam or column); width
b_{eff}	effective width
$b_{e,p}$	effective width for punching shear failure
c	projection width (base plates); equivalent width of the T-stub in compression
d	nominal bolt diameter; distance; clear depth of web (subscript b or c refers to beam or column)
d_0	hole diameter; diameter of the chord (structural hollow sections)
d_n	nominal diameter of the bolt shank
d_w	diameter of the washer, or the width across points of the head or nut, as relevant
e	distance from the centre of a fastener to the nearest edge (subscripts are defined for the particular case)
e_0	eccentricty
$f_{c,k}$	characteristic strength of the concrete foundation
$f_{j,d}$	concrete bearing strength
f_{pLT}	lateral torsional buckling strength of the plate
$f_{v,wd}$	design shear strength of a weld
f_u	material ultimate tensile strength (subscript f or w refers to flange or web, b or c refers to beam or column, d refers to bolt, p refers to plate)
f_y	material yield strength (subscript f or w refers to flange or web, b or c refers to beam or column, d refers to bolt, p refers to plate)
g	gap (in a lattice girder gap joint)
h	section depth (subscript b or c refers to beam or column); height
h_r	the distance from bolt-row r to the centre of compression
k	factor
k_1	coefficient
k_{eq}	equivalent stiffness coefficient

$k_{E;\theta}$	reduction factor for stiffness at temperature θ
k_i	stiffness coefficient of component i
k_j	concentration factor
k_s	reduction factor shape and size of the hole
$k_{y;\theta}$	reduction factor for the yield strength at temperature θ
k_{wc}	reduction factor for column web instability
l	lenght
l_{eff}	effective length (of a T-stub flange)
m	distance from the centre of a fastener to a fillet weld or to the radiuis of a rolled section fillet (measured to a distance into the fillet equal to 20% of its size, in both cases)
$m_{pl,Rd}$	design plastic moment of a plate per unit length
n	number of friction surfaces; effective edge distance (of a T-stub flange)
n_b	number of bolt rows (with two bolts per row)
p	uniformly distributed load; spacing between centres of fasteners (subscripts are defined for the particular case)
r	root radius of a rolled section (subscript b or c refers to beam or column)
s	length
t	thickness (subscript f or w refers to flange or web, b or c refers to beam or column, p refers to plate)
t_0	thickness of the chord
t_i	thickness of the braces, $i = 1,2$
u	ground settlement
w	horizontal spacing between lines of bolts in and end plate connection (gauge)
z	lever arm of the resultant tensile and compressive forces in the connection
z_{eq}	equivalent lever arm
xx	longitudinal axis of a member
yy	major axis of a cross section
zz	minor axis of a cross section

UPPER CASES

A	gross section area
A_0	original cross sectional area

A_{net}	net area of an angle
A_{nt}	net area subjected to tension
A_{nv}	net area subjected to shear
A_s	tensile stress area of a bolt
A_v	shear area (subscript b or c refers to beam or column)
A_w	throat section area
B_p	punching shear resistance of the bolt head and the nut
C	factor
$C_{f,d}$	coefficient of friction between base plate and grout layer
E	Young's modulus for steel material
F	force
F_b	tensile and compressive forces in the connection, statically equivalent to the beam end moment
F_e	external load
$F_{f,Rd}$	design friction resistance between base plate and grout layer
$F_{p,Cd}$	design preload force
$F_{t,Rd}$	design tension resistance of a bolt
$F_{T,Rd}$	design tension resistance of a T-stub flange
H	horizontal load
I	second moment of area (subscript b or c refers to beam or column)
K	buckling length coefficient
L	member length (subscript b or c refers to beam or column)
L_b	bolt elongation length
L_j	distance between the centres of the end fasteners in a joint
M	bending moment (subscript b or c refers to beam or column at the location of the joint)
M'	resistance per unit length of the T-stub flange in compression
N	axial force (subscript b or c refers to beam or column at the location of the joint)
$N_{net,Rd}$	design tension resistance of the net section at holes for fasteners
Q	prying force
P	axial compressive load
S	rotational joint stiffness
V	shear force (subscript b or c refers to beam or column at the location of the joint)
V_{eff}	design block tear resistance

V_n	shear force experienced by the column web panel
W	gravity load

GREEK LETTERS

α_b	coefficient (bolted joints)
α_{cr}	minimum force amplifier to reach the elastic critical buckling load
α_d	coefficient (bolted joints)
β	transformation parameter; general parameter
β_j	reduction factor
β_{LF}	reduction factor for long bolted joints
β_{LW}	reduction factor for long welded joints
β_w	correlation factor (for the design of welds)
ε	strain; parameter $\left(= \sqrt{235/f_y}\right)$
$\dot{\varepsilon}$	strain rate
δ	magnitude of the member deflection or local second-order effect
Δ	sway displacement or global second-order effect
Δ_i	deformation of spring i
ϕ	rotation
ϕ_{Cd}	design rotation capacity of a joint
ϕ_{pl}	plastic rotation capacity
γ	shear deformation
γ_c	partial safety factor for concrete
γ_F	partial safety factor for the loads (actions)
γ_M	partial safety factor for the resistance (strength function)
η	stiffness reduction factor
φ	angle
λ	slenderness
λ_{cr}	elastic critical load parameter (gravity loads)
λ_L	load parameter
λ_{ov}	overlap ratio
λ_p	plastic load parameter
$\bar{\lambda}$	reduced slenderness
μ	slip factor
θ	member rotation (subscript b or c refers to beam or column); temperature

Θ_i	angles between chords and brace members (hollow sections)
ρ	reduction factor allowing for plate buckling in the web
σ	normal stress
$\sigma_{com,Ed}$	maximum longitudinal compressive stress in a column due to axial force and bending
τ	shear stress
ω	reduction factor for interaction with shear
ψ	load combination factor

SUBSCRIPTS

1,2	brace members
a	angle cleat
app	approximate
b	beam; bolt, bearing
c	column; connection; compression
com	compression
cp	circular pattern
cr	critical
e	elastic
Ed	design value of effects of actions
eff	effective
eq	equivalent
f	flange
g	grout
ini	initial
j	joint
lim	limit
lo	lower
nc	non-circular pattern
ov	overlap.
p	plate; pin
pl	plastic conditions
$post\text{-}lim$	post-limit
r	bolt row
Rd	design value of resistance
s	slip

serv	SLS conditions
t	tension
u	ultimate
upp	upper
v	shear
w	weld; web
wp	web panel
X	extension of an end plate

ABBREVIATIONS

CHS	circular hollow section
E	elastic
P	plastic
RHS	rectangular hollow section
SHS	square hollow section
SLS	Service limit state(s)
ULS	Ultimate limit state(s)

Chapter 1

INTRODUCTION

1.1 GENERAL

1.1.1 Aims of the book

The aim of the present book is threefold:

- To provide designers with practical guidance and tools for the design of steel joints;
- To point out the importance of structural joints on the response of steel structures and to show how the actual behaviour of joints may be incorporated into the structural design and analysis process;
- To illustrate the possibilities of producing more economical structures using the new approaches offered in the European code of practice for the design of steel structures, Eurocode 3 (EN 1993) as far as structural joints are concerned.

The organisation of the book reflects the belief that, in addition to the sizing of the members (beams and columns), consideration should also be given to the joint characteristics throughout the design process. This approach, despite the novelty it may present to many designers, is shown to be relatively easy to integrate into everyday practice using currently available design tools.

Hence the present book addresses design methodology, structural analysis, joint behaviour and design checks, at different levels:

- Presentation and discussion of concepts;
- Practical guidance and design tools.

1.1.1.1 The traditional common way in which joints are modelled for the design of a frame

Generally speaking, the process of designing building structures has been up to now made up of the following successive steps:

- Frame modelling including the choice of rigid or pinned joints;
- Initial sizing of beams and columns;
- Evaluation of internal forces and moments (load effects) for Ultimate Limit States (ULS) and Serviceability Limit States (SLS);
- Design checks for ULS and SLS criteria for the structure and the constitutive beams and columns;
- Iteration on member sizes until all design checks are satisfactory;
- Design of joints to resist the relevant member end forces and moments (either those calculated, or the maximum ones able to be transmitted by the actual members); the design is carried out in accordance with the prior assumptions (frame modelling) on joint stiffness.

This approach was possible since designers were accustomed to considering the joints to be either pinned or rigid. In this way, the design of the joints became a separate task from the design of the members. Indeed, joint design was often performed at a later stage, either by other members of th design team or by another company.

Recognising that most joints have an actual behaviour which is intermediate between that of pinned and rigid joints, EN 1993 offers the possibility to account for this behaviour by opening up the way to what is presently known as the semi-continuous approach. This approach offers the potential for achieving better and more economical structures.

1.1.1.2 The semi-continuous approach

The rotational behaviour of actual joints is well recognised as being often intermediate between the two extreme situations, i.e. rigid or pinned. In sub-chapter 1.2, the difference between joints and connections will be introduced. For the time being, examples of joints between one beam and one column only will be used. Let us now consider the bending moments and the related rotations at a joint (Figure 1.1).

When all the different parts in the joint are sufficiently stiff (i.e. ideally infinitely stiff), the joint is rigid, and there is no difference between the respective rotations at the ends of the members connected at this joint (Figure 1.1a). The joint experiences a single global rigid-body rotation which is the nodal rotation in the commonly used analysis methods for framed structures.

Should the joint have only negligible stiffness, then the beam will behave just as a simply supported beam, whatever the behaviour of the other connected member(s) (Figure 1.1b). This is a pinned joint.

For intermediate cases (non-zero and finite stiffness), the moment transmitted will result in a difference ϕ between the absolute rotations of the two connected members (Figure 1.1c). The joint is classified as semi-rigid in these cases.

The simplest way for representing this concept is a rotational (spiral) spring between the ends of the two connected members. The rotational stiffness S_j of this spring is the parameter that links the transmitted moment M_j to the relative rotation ϕ, which is the difference between the absolute rotations of the two connected members.

When this rotational stiffness S_j is zero, or when it is relatively small, the joint falls back into the pinned joint class. In contrast, when the rotational stiffness S_j is infinite, or when it is relatively high, the joint falls into the rigid joint class. In all the intermediate cases, the joint belongs to the semi-rigid joint class.

For semi-rigid joints the loads will result in both a bending moment M_j and a relative rotation ϕ between the connected members. The moment and the relative rotation are related through a defined behaviour depending on the joint properties. This is illustrated in Figure 1.2 where, for the sake of simplicity, an elastic response of the joint is assumed in view of the structural analysis to be

3

(a) Rigid Joint   (b) Pinned joint   (c) Semi-rigid joint
Figure 1.1 – Classification of joints according to stiffness

performed (how to deal with non-linear behaviour situations will be addressed later on, especially in chapter 2).

It has to be understood that the effect, at the global analysis stage, of having semi-rigid joints instead of rigid or pinned joints is to modify not only the displacements, but also the distribution and magnitude of the internal forces throughout the structure.

As an example, the bending moment diagrams in a fixed-base simple portal frame subjected to a uniformly distributed load are given in Figure 1.3 for

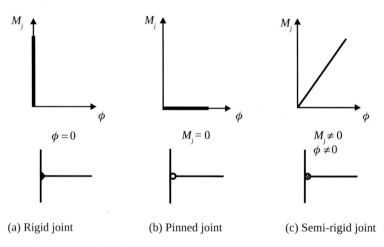

(a) Rigid joint   (b) Pinned joint   (c) Semi-rigid joint

Figure1.2 – Modelling of joints (case of elastic global analysis)

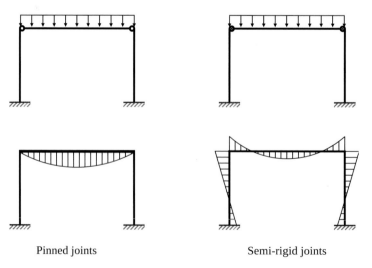

Pinned joints   Semi-rigid joints

Figure 1.3 – Elastic distribution of bending moments in a simple portal frame two situations, where the beam-to-column joints are respectively either pinned or semi-rigid. The same kind of consideration holds for deflections.

1.1.1.3 The merits of the semi-continuous approach

Both the EN 1993 requirements and the desire to model the behaviour of the structure in a more realistic way leads to the consideration of the semi-rigid behaviour when necessary.

Many designers would stop at that basic interpretation of EN 1993 and hence would be reluctant to confront the implied additional computational effort involved. Obviously a crude way to deal with this new burden will be for them to design joints that will actually continue to be classified as being pinned or fully rigid. However, those properties will have to be proven at the end of the design process; in addition, such joints will certainly be uneconomical in a number of situations.

It shall be noted that the concept of rigid and pinned joints still exists in EN 1993. It is accepted that a joint which is almost rigid or, on the contrary, almost pinned, may still be considered as being truly rigid or truly pinned in the design process. How to judge whether a joint can be considered as rigid, semi-rigid or pinned depends on the comparison between the joint stiffness and the beam stiffness, which depends on the second moment of area and length of the beam.

The designer is strongly encouraged to go beyond this "all or nothing" attitude. Nowadays, it is possible, and therefore of interest, to consider the benefits to be gained from the semi-rigid behaviour of joints. Those benefits can be brought in two ways:

1. The designer decides to continue with the practice of assuming – sometimes erroneously – that joints are either pinned or fully rigid. However, EN 1993 requires that proper consideration be given to the influence that the actual behaviour of the joints has on the global behaviour of the structure, i.e. on the accuracy with which the distribution of forces and moments and the displacements have been determined. This may not prove to be easy when the joints are designed at a late stage in the design process since some iteration between global analysis and design checking may be required. Nevertheless, the following situations can be foreseen:

5

- For a joint to be assumed rigid, it is common practice to introduce, for instance, column web stiffeners in a beam-to-column joint. EN 1993 now provides the means to check whether such stiffeners are really necessary for the joint to be both rigid and have sufficient resistance. There are practical cases where they are not needed, thus permitting the adoption of a more economical joint design;
- When joints assumed to be pinned are later found to have fairly significant stiffness (i.e. to be semi-rigid), the designer may be in a position to reduce beam sizes. This is simply because the moments carried out by the joints reduce the span moments and deflections in the beams.

2. The designer decides to give consideration, at the preliminary design stage, not only to the properties of the members but also to those of the joints. It will be shown that this new approach is not at all incompatible with the frequent separation of the design tasks between those who have the responsibility for conceiving the structure and carrying out the global analysis and those who have the responsibility for designing the joints. Indeed, both tasks are very often performed by different people, indeed, or by different companies, depending on national or local industrial habits. Adopting this novel approach towards design requires a good understanding of the balance between, the costs and the complexity of joints and, the optimisation of the structural behaviour and performance through the more accurate consideration of joint behaviour for the design as a whole. Two examples are given to illustrate this:

   - It was mentioned previously that it is possible in some situations to eliminate stiffeners, and therefore to reduce costs. Despite the reduction in its stiffness and, possibly, in its strength, the joint can still be classified as rigid and found to have sufficient strength. This is possible for industrial portal frames with rafter-to-column haunch joints, in particular, but other cases can be envisaged;
   - In a more general way, it is worth considering the effect of adjusting the joint stiffness so as to strike the best balance between the cost of the joints, and the cost of the beams and columns. For instance, for braced frames, the use of semi-rigid

joints, which are probably more costly than the pinned joints, enables reducing the beam sizes. For unbraced frames, the use of less costly semi-rigid joints, instead of the rigid joints, leads to increased column sizes and possibly beam sizes.

Of course the task may seem difficult, and this is why the present book, in chapter 9, is aimed at providing the designer with a set of useful design strategies. The whole philosophy could be summarised as "*As you must do it, so better make the best of it*".

Thus, EN 1993 now offers the designer the choice between a conservative attitude, where however something may often be gained, and an innovative attitude, where the economic benefits may best be sought.

It is important to stress the high level of similarity that exists between response of member cross sections and that of structural joints. This topic is addressed in the following section.

1.1.1.4 A parallel between member cross sections and joints

Member cross section behaviour may be considered by means of a moment-rotation curve, $M - \theta$, for a simply supported beam loaded at mid span (M here is the bending-moment at mid-span, and θ is the sum of rotations at the span ends). Joint behaviour will be considered through a similar relationship, but with $M = M_j$ being the bending moment transmitted by the joint, and $\theta = \phi$ being the relative rotation between the connected members. Those relationships have a similar shape as illustrated in Figure 1.4.

According to EN 1993, member cross sections are divided into four classes, especially according to their varying ability to resist local instability, when partially or totally subject to compression, and the consequences this may have on the possibility for plastic redistribution. Therefore, their resistance ranges from the full plastic resistance (class 1 and 2) to the elastic resistance (class 3) or the sub-elastic resistance (class 4).

The specific class to which a cross section belongs defines the assumptions on:

  – The behaviour to be idealised for global analysis (i.e. class 1 will allow the formation of a plastic hinge and the redistribution of internal forces in the frame as loads are increased up to or beyond the design loads);

7

    – The behaviour to be taken into account for local design checks (i.e. class 4 will imply that the resistance of the cross section is based on the properties of a relevant effective cross section rather than those of the gross cross section).

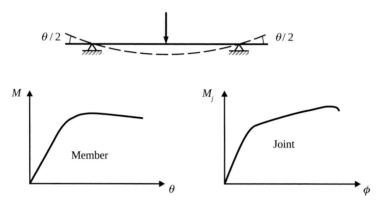

Figure 1.4 – Moment vs. rotation characteristics for member cross section and joint

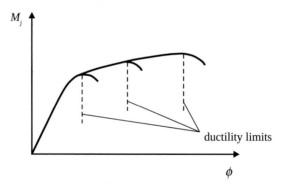

Figure 1.5 – Ductility or rotation capacity in joints

      The classification of a cross section is based on the width-to-thickness ratio of the steel component walls of the section. Ductility is directly related to the amount of rotation during which the design bending resistance will be sustained. This results in a so-called rotation capacity concept.

      In a similar manner, joints are classified in terms of ductility or rotation capacity, see Figure 1.5. This classification is a measure of their ability to resist premature local instability, and premature brittle failure (especially due to bolt or weld failure) with due consequences on the type of global analysis allowed.

      The practical interest of such a classification for joints is to check whether an elastoplastic global analysis may be conducted up to the formation of a plastic collapse mechanism in the structure, which implies the formation of hinges in some of the joints.

As will be shown, this classification of joints by ductility, although not explicitly stated in EN 1993, may be defined from the geometric and mechanical properties of the joint components (bolts, welds, plate thickness, etc.). Joints may therefore be classified according to both stiffness, and ductility.

Additionally, joints may be classified according to their strength, as full-strength or partial-strength, according to their resistance compared to the resistance of the connected members. For elastic design, the use of partial-strength joints is well understood. When plastic design is used, the main use of this classification is to anticipate the possible need to allow a plastic hinge to form in the joint for global analysis. In order to allow increase of loads, a partial-strength joint may be required to act as a plastic hinge from the moment its bending resistance is reached. In that case, the joint must also have sufficient ductility (plastic rotation capacity).

The final parallel to be stressed between joints and members is that the same kind of link exists between the global analysis stage and the ultimate limit states design checking stage. The latter has to extend to all aspects that were not implicitly or explicitly taken into account at the global analysis stage. Generally speaking, it can be stated that the more sophisticated the global analysis is, the simpler the ULS design checks required.

The choice of global analysis will thus depend not only on what is required by EN 1993 but also on personal choices, depending on specific situations, available software tools, etc. A particular choice means striking a balance between the amount of effort devoted to global analysis and the amount of effort required for the check of remaining ULS, see Figure 1.6.

Figure 1.6 – Schematic of the proportion of effort for global analysis and for ULS checks

---

UK Specific Comment

UK practice normally considers only two main forms of joint modelling, simple and continuous. Ideally frames designed on the basis of simple joint modelling should use joints between members that possess negligible rotational stiffness, and transmit the beam reactions in shear into the columns without developing significant moments. These joints may be treated as perfect pins. Members can then be designed in isolation, either as (predominantly) axially loaded columns, or as simply supported beams. Joints designed within the principles of continuous modelling are capable of transmitting significant moments and are able to maintain the original angle between adjacent members virtually unchanged. This form of construction of steel frames is particularly advantageous when beam deflections are critical or if bracing systems are not possible.

To comply with one or the other design assumption, steelwork connections are detailed as simple or moment-resisting joints. Because joints in frame structures are often treated by adopting a degree of standardisation, the designer usually details the connections in accordance with the principles in the series of SCI Green Books, which present this material in the form of step-by-step design procedures and tables covering standard arrangements.

Other approximate analysis methods continue to be used among British designers, the most representative being the wind-moment method for the design of unbraced multi-storey steel frames. The method assumes that (i) under gravity loads the beam-to-column joints act as pinned connections, and (ii) under horizontal wind loads these joints are rigid, although it lacks transparency. This method can be seen as a manifestation of the semi-continuous principles. Wind moment connection details can be found in SCI (1995a). With the now widespread use of software in design offices for rigorous second-order analysis, it is doubtful if the use of the wind-moment method should be advocated any longer, other than for initial sizing.

10

1.1.2 Brief description of the contents of the book

This book is divided into three main parts:

- The first part covered by chapter 2 is a recall of the available methods of structural analysis and design, but with a special focus on the integration of the joint response into the whole design process;

- The second part includes chapter 3 to chapter 8. It concerns the characterisation of the joint properties in terms of resistance, stiffness, ductility. Chapter 3 to chapter 7 concentrate on the response of joints under static loading while chapter 8 provides information about the extension of the previously described joint characterisation methods to fire, seismic and fatigue loading;
- In the third part, covered by chapter 9, design opportunities to optimise the structural frames and the constitutive joints are presented, and guidelines are provided on how to implement them into daily practice.

Practical design recommendations and tools are presented throughout the whole book. This is intended to help the reader to "materialise" the possibly new concepts and the way to "operate" them in their design activities. In order to achieve a full consistency with the European norms, the notations adopted in the book are basically those suggested in EN 1993. The design rules implemented in the codes are not all repeated in the present book, which is aimed at commenting the joint related aspects of EN 1993, but certainly not at "replacing" the normative documents.

Finally, it has to be mentioned that only one load case is generally considered in the book, when referring to the design of a specific joint or frame, whereas, in actual projects, several load combination cases have obviously to be dealt with.

1.1.3 Types of structural systems and joints covered

No particular type of structure is excluded but the book is focusing more particularly on steel building frames. As far as joints are concerned, the configuration and the loading conditions addressed in this book are, strictly speaking, those directly covered by part 1-8 of EN 1993, EN 1993-1-8 (CEN, 2005c); however, as the Eurocode design principles are also valid for many other types of joints and loading conditions, the scope the present book extends to the following:

- Member cross sections: open H or I rolled and welded sections, as well as tubular sections;
- Connections with mechanical fasteners;
- Welded connections;

    – Beam-to-column joints, beam-to-beam joints, and column or beam splices: simple and moment-resisting joints with welded or bolted steel connections (with or without haunch) and using various connecting elements (end plates, angles, fin plates) and stiffening systems;

    – Column bases: bolted end plate connections;

    – Lattice girder joints;

    – Joints under static and non-static loading, including fire, seismic and fatigue loading.

1.1.4 Basis of design

EN 1993-1-8 is one of the twelve constitutive parts of EN 1993. Accordingly, its basis for design is fully in line with EN 1993, in terms of basic concepts, basic variables, ULS, SLS, durability and sustainability. Reference will therefore be done here to the introductory chapter of the ECCS Design Manual "Design of steel structures" (Simões da Silva *et al*, 2010) where the here-above mentioned aspects are extensively discussed.

1.2   DEFINITIONS

One of the advantages of the steel construction lies in the prefabrication of steel pieces. These pieces, according to the needs or the transportation requirements, will have to be assembled either on site or directly in the workshop. The connections resulting from this fabrication and erection process will sometimes be there just for constructional reasons; but in many cases, they will have to transfer forces between the connected pieces and so play an important structural role. One speaks then about "structural connections, which are a particular focus of this book.

In the past, forging was used as a convenient way to assemble steel pieces together. This was achieved through the melting of the steel at the interface between the connected pieces. Nowadays, in modern constructions, use is made of welds and mechanical fasteners, including various bolt types (e.g. TC bolts, injection bolts, flow drill bolting), pins, or nails. Beside the connectors, a great variety of configurations is also required in daily practice, sometimes as a result of the architectural demand or for economic reasons. This leads to a number of situations to which the practitioner is likely to be faced. Moreover, various loading situations (e.g. static, dynamic, seismic, fire) and structural systems (e.g. buildings, bridges, towers) are also to be considered.

In this book, an exhaustive consideration of all the existing connection and joint configurations is obviously not practical; therefore the authors have tried to deliver the most updated and complete information in the domain by structuring chapter 3 to chapter 8 as follows:

- Chapter 3: Connections with mechanical fasteners;
- Chapter 4: Welded connections;
- Chapter 5: Simple joints;
- Chapter 6: Moment-resisting joints;
- Chapter 7: Lattice girder joints;
- Chapter 8: Joints under various loading situations.

Chapter 3 and 4 refer to basic joints where mainly tension and/or shear forces are to be transferred between the connected pieces. The most important design property is their level of resistance.

Chapter 5 is devoted to connections aimed at transferring shear forces, possibly together with axial forces, between two elements, but additionally allow free rotation between these elements. The word "simple" refers to this specific ability.

Chapter 6 concentrates on moment-resisting joints where a combination of axial/shear forces and moments are to be carried over by the joint. Their rotational stiffness plays an important role in terms of global structural response. In the case of plastic frame analysis, their rotational capacity will also have to be assessed.

In chapter 7, lattice girder joints are discussed, and, in chapter 8, the extension of the static design rules to fire, fatigue or earthquake situations will be discussed.

The design of the basic connections (covered in chapters 3 and 4) and of lattice girder joints according to the Eurocodes is quite similar to what practitioners have been used to apply in the last decades at the national level while new concepts are now available for the design of simple and moment-resisting joints. The next pages will allow the designer to familiarise with important definitions used in the book.

13

1.2.1 Joint properties

The specific case of building frames is addressed below, just for the sake of illustration.

Building frames consist of beams and columns, usually made of H, I or tubular shapes that are assembled together by means of connections. These connections are between two beams, two columns, a beam and a column, or a column and the foundation (Figure 1.7).

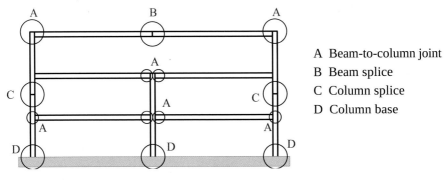

A Beam-to-column joint
B Beam splice
C Column splice
D Column base

Figure 1.7 – Different types of joints in a building frame

(a) Single sided joint configuration   (b) Double sided joint configuration

Figure 1.8 – Joints and connections

A *connection* is defined as the set of the physical components which mechanically fasten the connected elements. The connection is assumed to be concentrated at the location where the fastening action occurs, for instance at the beam end/column interface in a beam-to-column joint. When the connection as well as the corresponding zone of interaction between the connected members are considered together, the wording *joint* is used (Figure 1.8a).

Depending on the number of in-plane elements connected together, single-sided and double-sided joint configurations are defined (Figure 1.8). In a double-sided configuration (Figure 1.8b), two joints – left and right – have to be considered.

The definitions illustrated in Figure 1.8 are valid for other joint configurations and connection types.

As explained previously, the joints, which are traditionally considered as rigid or pinned, and designed accordingly, possess, in reality, their own degree of flexibility resulting from the deformation of all the constitutive components. Section 1.2.2 describes the main *sources of joint deformation*. Sub-chapter 2.2 provides information on how to *model* the joints in view of frame analysis. This modelling depends on the level of joint flexibility. In sub-chapter 2.3, the way in which the shape of the non-linear joint deformation curves may be *idealised* is given. Section 1.6.2 refers to the *component method* as a general tool for the prediction of the main joint mechanical properties in bending. The concept of *joint classification* is introduced in sub-chapter 2.4. Finally, it is commented on the *ductility classes* of joints in sub-chapter 2.5. This aspect, expressed in terms of rotational capacity, will be of particular importance in cases where a plastic structural analysis is carried out.

1.2.2 Sources of joint deformation

As stated in sub-chapter 1.1, the rotational behaviour of the joints may affect the local and/or global structural response of the frames. In this sub-chapter, the sources of rotational deformation are identified for beam-to-column joints, splices and column bases.

It is worth mentioning that the rotational stiffness, the resistance and the rotation capacity are likely to be affected by the shear force and/or the axial force acting in the joint. These shear and axial forces may obviously have contributions to the shear and axial deformation within the joint. However, it is known that these contributions do not affect significantly the frame response. Therefore, the shear and axial responses of the joints, in terms of rotational deformation, are usually neglected in applications as buildings.

1.2.2.1 Beam-to-column joints

1.2.2.1.1 Major axis joints

In a major axis beam-to-column joint, different sources of deformation can be identified. For the particular case of a single-sided joint (Figure 1.9a and Figure 1.10a), these are:

- *The deformation of the connection* that includes the deformation of the connection elements: column flange, bolts, end plate or angles, and the load-introduction deformation of the column web resulting from the transverse shortening and elongation of the column web under the compressive and tensile forces F_b acting on the column web. The couple of forces F_b are statically equivalent to the moment M_b at the beam end. These deformations result in a relative rotation ϕ_c between the beam and column axes; this rotation, which is equal to $\theta_b - \theta_c$ (see Figure 1.9a), results in a flexural deformation curve $M_b - \phi_c$.
- *The shear deformation of the column web panel* associated with the shear force V_{wp} acting in this panel. It leads to a relative rotation γ between the beam and column axes; this rotation makes it possible to establish a shear deformation curve $V_{wp} - \gamma$.

The deformation curve of a connection may obviously be influenced by the axial and shear forces possibly acting in the connected beam.

Similar definitions apply to double-sided joint configurations (Figure 1.9b and Figure 1.10 b). For such configurations, two connections and a sheared web panel, forming two joints, must be considered.

To summarise, the main sources of deformation which must be taken into account in a beam-to-column major axis joint are:

Single-sided joint configuration:

- The connection deformation $M_b - \phi_c$ characteristic;
- The column web panel shear deformation $V_{wp} - \gamma$ characteristic.

Double-sided joint configuration:

- The right hand side connection deformation $M_{b1} - \phi_{c1}$ characteristic;
- The left hand side connection deformation $M_{b2} - \phi_{c2}$ characteristic;
- The column web panel shear deformation $V_{wp} - \gamma$ characteristic.

(a) Single sided joint configuration   (b) Double sided joint configuration

Figure 1.9 – Sources of joint deformation

The deformation of the connection elements and load-introduction is only due to the couple of forces transferred by the flanges of the beam, equivalent to the beam end moment M_b. The shear deformation of the column web panel results from the combined action of these equal but opposite forces and of the shear forces in the column at the level of the beam flanges. Equilibrium equations of the web panel provide the shear force V_{wp} (for the sign convention, see Figure 1.10, where the directions indicated by the arrows are positive):

$$V_{wp} = \frac{M_{b1} - M_{c2}}{z} - \frac{V_{c1} - V_{c2}}{2} \tag{1.1}$$

If the beneficial effect of the shear force in the column is neglected, Eq. (1.1) simplifies as follows:

$$V_{wp} = \frac{M_{b1} - M_{b2}}{z} \tag{1.2}$$

In both formulae, z is the lever arm of the resultant tensile and compressive forces in the connection(s). Chapter 6 explains how to derive the value of z.

1.2.2.1.2 Minor axis joints

A similar distinction between *web panel* and *connection* must also be made for a minor axis joint (Figure 1.11). The column web exhibits a so-called out-of-plane deformation while the connection deforms in bending, as in a major axis joint. However, no load-introduction deformation is involved.

In the double-sided joint configuration, the out-of-plane deformation of the column web depends on the bending moments experienced by the right and left connections (see Figure 1.12):

17

$$\Delta M_b = M_{b1} - M_{b2} \tag{1.3}$$

For a single-sided joint configuration (Figure 1.11), ΔM_b is M_b.

1.2.2.1.3 Joints with beams on both major and minor column axes

A three-dimensional (3-D) joint is characterised by the presence of beams connected to both the column flange(s) and web (Figure 1.13). In such joints, a shear deformation (see 1.2.2.1.1) and an out-of-plane deformation (see 1.2.2.1.2) of the column web develop simultaneously.

The loading of the web panel appears therefore as the superimposition of the shear loading given by Eq. (1.1) or (1.2) and the out-of-plane loading given by Eq. (1.3).

The joint configuration of Figure 1.13 involves two beams only; configurations with three or four beams can also be found.

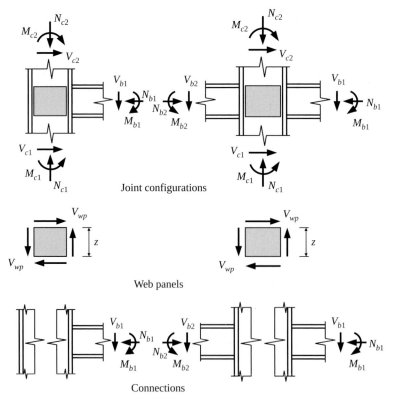

Joint configurations

Web panels

Connections

(a) Single sided joint configuration   (b) Double sided joint configuration

Figure 1.10 – Loading of the web panel and the connections

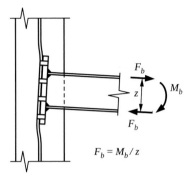

Figure 1.11 – Deformation of a minor axis joint

Figure 1.12 – Loading of a double-sided minor axis joint

19

Figure 1.13 – Example of a 3-D joint

1.2.3 Beam splices and column splices

The sources of deformation in a beam splice (Figure 1.14) or in a column splice (Figure 1.15) are less than in a beam-to-column joint; indeed they are concerned with connections only. The deformation is depicted by the sole curve $M_b - \phi_c$.

This curve corresponds to the deformation of the whole joint, i.e. the two constituent connections (left connection and right one in a beam splice, upper connection and lower one in a column splice).

In a column splice where the compressive force is predominant, the axial force affects the mechanical properties of the joint significantly, i.e. its rotational stiffness, its strength and its rotation capacity. The influence, on the global frame response, of the axial deformation of splices is however limited; therefore it may be neglected provided that specific requirements given in sub-chapter 6.7 are fulfilled.

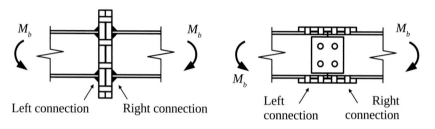

Figure 1.14 – Deformation of a beam splice

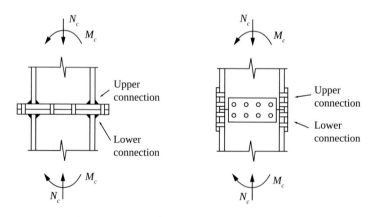

Figure 1.15 – Deformation of a column splice

1.2.4 Beam-to-beam joints

The deformation of a beam-to-beam joint (Figure 1.16) is similar to that of a minor axis beam-to-column joint; the loadings and the sources of deformation are similar to those expressed in 1.2.2.1.2 and can therefore be identified.

Figure 1.16 – Deformation of a beam-to-beam joint

1.2.5 Column bases

In a column base, two connection deformabilities need to be distinguished (Figure 1.17):

- The deformation of the connection between the column and the concrete foundation (*column-to-concrete connection*);
- The deformation of the connection between the concrete foundation and the soil (*concrete-to-soil connection*).

For the column-to-concrete connection, the bending behaviour is represented by a $M_c - \phi$ curve, the shape of which is influenced by the ratio of the bending moment to the axial load at the bottom of the column. For the connection between the concrete foundation and the soil, two basic deformation curves are identified:

- A $N_c - u$ curve which mainly corresponds to the soil settlement, u, due to the axial compressive force in the column, N_c; in contrast with the other types of joint, this deformation curve may have a significant effect on the frame behaviour;
- A $M_c - \phi$ curve characterising the rotation of the concrete block in the soil.

As for all the other joints described above, the deformation due to shear forces in the column may be neglected in the case of column bases.

The column-to-concrete connection and concrete-to-soil connection $M_c - \phi$ characteristics are combined in order to derive the rotational stiffness at the bottom of the column and conduct the frame analysis and design accordingly.

Similar deformation sources exist in column bases subjected to biaxial bending and axial force. The connection $M_c - \phi$ characteristics are then defined for both the major and the minor axes.

1.2.6 Hollow section joints

Design rules in EN 1993-1-8 for hollow section joints were originally developed and published as design recommendations by the International Committee for Research and Technical Support for Hollow Section Structures (CIDECT). Chapter 7 of EN 1993-1-8 gives application rules to determine the static design resistances of joints in lattice structures.

Design formulae for hollow section joints are based on semi-empirical investigations in which analytical models were fitted with test results. The complex geometry of the joints, local influences of the corners of rectangular sections and residual stresses, for instance due to welding, lead to non-uniform stress distributions. Strain hardening and membrane effects also influence the local structural behaviour. Simplified analytical models which consider the most relevant parameters were developed. In contrast to the design models for open section joints, where the so-called component method (see section 1.6.2) is used, the design of hollow section joints is based on the study of failure modes. Note that the terminology is not fully consistent here. In the previous sections, the terms *joint configuration, joint* and *connection* respectively are defined. For the design of hollow section joints, different types of joints are covered in EN 1993, see Table 1.1. However, here the word "joint" means *joint* and *joint configuration* at the same time. Hence, the resistance of a joint is also the resistance of the whole node, i.e. joint configuration, typically expressed by the resistances of the braces, see Figure. 1.18.

1.3   MATERIAL CHOICE

The design provisions given in the Eurocodes are only valid if material or products comply with the reference standards given in the appropriate parts of the Eurocodes. Both, in EN 1993-1-1 (CEN, 2005a) and in EN 1993-1-8 (CEN, 2005c), material and product standards are listed in sections 1.2. EN 1993-1-1, as a master document for the other parts of EN 1993, covers the

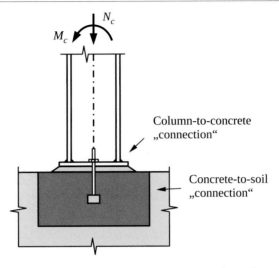

Figure 1.17 – The two connections in a column base

design of steel structures fabricated from steel material conforming to the steel grades listed in Table 3.1 of EN 1993-1-1. According to this table, also for EN 1993-1-8, only steel grades from S235 to S460 are covered. For hot rolled structural steel (I or H sections, and welded build-up sections), reference is made to EN 10025 (CEN, 2004d). For structural hollow sections, reference is made to EN 10210 (CEN, 2006c and 2006d) for hot finished sections and to EN 10219 (CEN, 2006a and 2006b) for cold formed sections. Especially with regard to the design of joints where, in many cases, a local plastic redistribution of stresses will be assumed to determine the resistance of a joint, sufficient ductility of the material is required. Steel conforming to steel grades listed in Table 3.1 of EN 1993-1-1 can be assumed to have sufficient ductility.

Due to the fabrication process, the nominal values of the yield strength f_y and also of the ultimate strength f_u depend on the thickness of the elements. It should be noted that different values for the relation between the material thickness and the material strength are given in the delivery condition speci fied in the product standards (for example, EN 10025) and in EN 1993. EN 1993 allows to define the strength values either by adopting the values directly from the product standards, or by using simplified values given in Table 3.1 of EN 1993-1-1. The user should check the National Annex where a choice could have been made.

23

Table 1.1 – Types of hollow section joints in lattice structures

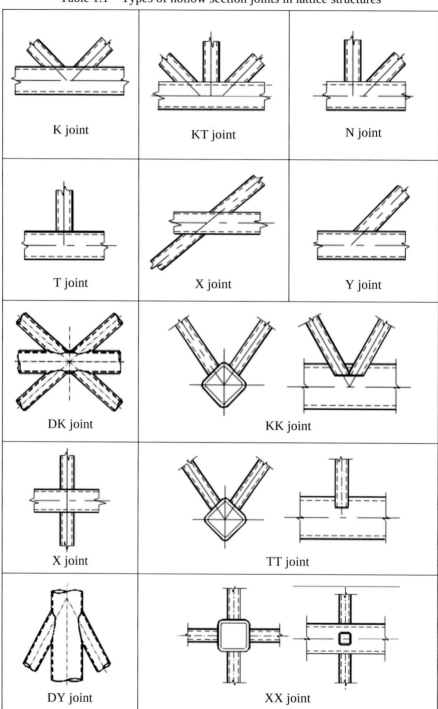

K joint	KT joint	N joint
T joint	X joint	Y joint
DK joint	KK joint	
X joint	TT joint	
DY joint	XX joint	

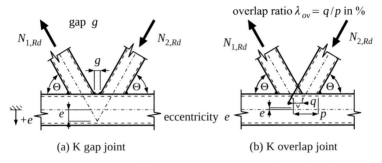

(a) K gap joint   (b) K overlap joint

Figure 1.18 – Definition of design resistance, gap and overlap of a K joint

To avoid brittle fracture of tension elements, the material should have sufficient facture toughness. EN 1993-1-10 (CEN, 2005e) provides rules to select the appropriate steel grade. Several aspects should be taken into account for the choice of material:

– Steel material properties: the yield strength depending on the material thickness $f_y(t)$ and the toughness quality, which is expressed in terms of T_{27J} or T_{40J};

– Member characteristics: the member shape, detail and element thickness (t), stress concentrations according to the details in EN 1993-1-9 (CEN, 2005d) as well as fabrication flaws (for example through-thickness);

– Design situations: design value at lowest member temperature as well as maximum stresses for permanent and imposed actions derived from the design residual stress should be considered. If relevant, assumptions for crack growth for fatigue loading during an inspection interval, the strain rate for accidental actions and the degree of cold forming.

In steel assemblies where the plate is loaded in tension perpendicular to its plane, i.e. in the through-thickness direction, the material must have adequate through-thickness properties to prevent lamellar tearing, see Figure 1.19. EN 10164 (CEN, 2004e) defined quality classes in terms of Z-values.

EN 1993-1-10 gives guidance on the quality class to be used where steel with improved through-thickness properties is necessary. This is for example the case of welded beam-to-column joints (where the column flange is loaded in its through-thickness direction) or beam-to-column joints with bolted end plates (where the end plate is loaded in its through-thickness direction).

Finally, for steel grades ranging from S460 to S690 (S700) reference is made to EN 1993-1-12 (CEN, 2007a).

Figure 1.19 – Lamellar tearing

UK Specific Comment

Although S460 is attracting growing interest, S355 is still the dominant grarsde of steel used in the UK.

1.4 FABRICATION AND ERECTION

The design process and the fabrication and erection of a structure are not independent. In many cases, and in particular for the joints, the design procedures are directly dependent on how a structure is built and vice versa. Eurocodes, strictly speaking, are design codes, or even verification codes, i.e. they provide the designer with rules to verify the resistance, stability, deformations and ductility requirements. But no guidelines related to fabrication and erection are given. For these specific aspects, the engineer must refer to the fabrication and erection, the so-called "execution standard", EN 1090-2 (CEN, 2011). This standard gives quite a number of detailed requirements to be fulfilled.

For bolted and welded joints, the link between design rules and fabrication/execution rules are particularly strong. For instance, the design of a bolt in bearing depends on the diameter of the bolt, but also on the diameter of the bolt hole. In practice, the hole clearance, i.e. the difference of diameter between the hole and the bolt diameter, is not specified in EN 1993-1-8, but the design rules provided there are only valid if the hole is drilled in accordance with EN 1090-2 provisions. If it is not the case, then the design rule proposed in EN 1993-1-8 cannot be applied.

Even if it is not explicitly specified in EN 1993-1-8, most of the design rules are only valid if the structural elements comply with the regulations specified in the execution/fabrication/erection standard EN 1090-2. Requirements are given for imperfections, geometric tolerances, e.g. tolerances on hole diameter for bolts and pins, methods of tightening of preloaded bolts, preparation of contact surfaces in slip resistant connections, welding plan and preparation and execution of welding, etc.

In this book, not specific focus is given to the fabrication and erection aspects. To know more about that, reference should be made to commentaries and background documentation of the execution standard EN 1090-2. But for sure all the design rules provided in this book may only be applied under the condition that the requirements of EN 1090-2 are satisfied. Similarly, all connected or connecting elements (bolts, welding materials, steel for end plates, etc.) are assumed to respect the EN product norms listed on the first pages of EN 1993-1-8.

1.5   COSTS

The price of a steel structure depends significantly on the fabrication, transportation and erection costs in countries where labour costs are the dominant factor. Hence, in order to design steel structures in an economical manner, special attention should be given to the joints which are known to be expensive. However, the joint detailing must also be chosen in such a way that the actual behaviour of the joint, i.e. its structural properties (strength, stiffness and ductility), is in line with the assumptions made when the joints have been modelled for the global frame analysis. In chapter 9, some strategies are presented which allow to minimise the effects of the joints on the global cost of steel structures.

1.6   APPLICATION OF THE "STATIC APPROACH"

The design of joints may, as for any other cross section, be performed on an elastic or a plastic basis. In a pure elastic approach, the joints should be designed in such a way that the generalised Von Mises stress does not exceed the elastic strength of the constitutive materials. For that purpose, and because of the geometric complexity of the connection elements, a refined stress analysis will be required, often requiring sophisticated numerical approaches such as finite element analysis. In this process, the presence of residual stresses or of any other set of self-equilibrated stresses which could, for instance, result from lack-of-fit, is usually neglected.

Steel usually exhibits significant ductility. The designer may therefore profit from this ability to deform steel plastically and so to develop "plastic" design approaches in which local stress plastic redistributions in the joint elements are allowed.

The use of plastic design approaches in steel construction is explicitly allowed in EN 1993. For joints and connections, reference is made to clause 2.5(1) of EN 1993-1-8. Obviously, limitations to the use of such plastic approaches exist. They all relate to the possible lack of ductility of the steel material, on the one hand, and of some connections elements like bolts, welds, reinforced concrete slab in tension or concrete in compression, on the other hand. As far as steel material itself is concerned, the use of normalised steels according to European standards, e.g. EN 10025, guarantees a sufficient material ductility. For potentially non ductile joint elements, EN 1993-1-8 introduces specific requirements which will be detailed later on when describing the design methods in dedicating chapters.

Clause 2.5(1) of EN 1993-1-8 is simply paraphrasing the so-called "static" theorem of the limit analysis. The application of this theorem to cross sections, for instance in bending, is well known as it leads to the concept of bi-rectangular stress patterns and to the notion of plastic moment resistance used for Class 1/2 cross sections (as no limit of ductility linked to plate buckling phenomena prevents these sections from developing a full plastic resistance). For connections and joints, the implementation of the static theorem is probably less direct, but this does not at all prevent designers from using it.

The static theorem requires first the determination of an internal statically admissible distribution of stresses (for cross sections) or forces (for connections and joints), i.e. of a set of stresses or forces in equilibrium with the external forces acting on the cross section or on the joint/connection and resulting from the global structural analysis. The second requirement is to be plastically admissible, which means that the plastic resistance and ductility criteria have to be met in the cross section or joint/connection.

A large number of statically and plastically admissible distributions may exist. In most cases, these will not respect the "kinematically admissible" criterion; in fact, only the actual distribution has the ability to satisfy the three requirements: equilibrium, plasticity and compatibility of displacements. But in fact, this is not a problem as long as sufficient local deformation capacity (ductility) is available at the places where plasticity develops in the cross section or in the joint/connection. As long as this last condition is fulfilled, the static theorem ensures that the predicted resistance of the cross section or joint/connection will be lower than the actual one (and therefore on the safe side). The closer the assumed distribution is to the actual one, the closer the estimated resistance will be to the actual resistance.

1.6.1 Component approach

1.6.1.1 General

The characterisation of the response of the joints in terms of stiffness, resistance and ductility is a key aspect for design purposes. From this point of view, three main approaches may be followed:

- Experimental;
- Numerical;
- Analytical.

The most practical one for the designer is the analytical approach. Analytical procedures enable a prediction of the joint response based on the knowledge of the mechanical and geometric properties of the so-called "joint components".

In this section the *component method* is introduced as a general analytical procedure. It applies to any type of steel joints, whatever the geometric configuration, the type of loading (axial force and/or bending moment) and the type of member cross sections.

The method is nowadays widely recognised, and particularly in EN 1993, as a general and convenient procedure to evaluate the mechanical properties of joints subjected to various loading situations, including static and dynamic loading conditions, fire, and earthquake.

1.6.1.2 Introduction to the component method

A joint is generally considered as a whole and studied accordingly; the originality of the component method is to consider any joint as a set of individual basic components. For the particular joint shown in Figure 1.9a (joint with an extended end plate connection mainly subject to bending), the relevant components (i.e. zones of transfer of internal forces) are the following:

    – Column web in compression;
    – Beam flange and web in compression;
    – Column web in tension;
    – Column flange in bending;
    – Bolts in tension;
    – End plate in bending;
    – Beam web in tension;
    – Column web panel in shear.

Each of these basic components possesses its own strength and stiffness either in tension, compression or shear. The column web is subjected a combination of compression, tension and shear. This coexistence of several components within the same joint element can obviously lead to stress interactions that are likely to decrease the resistance of the individual basic components.

The application of the component method requires the following steps:

1. *identification* of the active components in the joint being considered;

2. *evaluation of the stiffness and/or resistance characteristics* for each individual basic component (specific characteristics – initial stiffness, design resistance, deformation capacity – or whole deformation curve);

3. *assembly* of all the constituent components and evaluation of the stiffness and/or resistance characteristics of the whole joint (specific characteristics - initial stiffness, design resistance, rotation capacity - or whole deformation curve).

The assembly procedure is the step where the mechanical properties of the whole joint are derived from those of all the individual constituent components. That requires, according to the static theorem introduced in section 1.6.1, to define how the external forces acting on the joint distribute into internal forces acting on the components in a way that satisfies equilibrium and respects the behaviour of the components.

Guidelines on how to apply the component method for the evaluation of the initial stiffness and the design moment resistance of the joints are given in EN 1993-1-8; the aspects of ductility are also addressed. The application of the component method requires a sufficient knowledge of the behaviour of the basic components. Those covered for static loading by EN 1993-1-8 are listed in Table 1.2. The combination of these components covers a wide range of joint configurations and should be largely sufficient to satisfy the needs of practitioners. Examples of such joints are given in Figure 1.20.

Table 1.2 – List of components covered in EN 1993-1-8

No	Component	
1	Column web panel in shear	
2	Column web in transverse compression	
3	Column web in transverse tension	
4	Column flange in bending	
5	End plate in bending	

Table 1.2 – List of components covered by Eurocode 3 Part 1-8, (continuation)

No	Component	
6	Flange cleat in bending	$F_{t,Ed}$
7	Beam or column flange and web in compression	$F_{c,Ed}$
8	Beam web in tension	$F_{t,Ed}$
9	Plate in tension or compression	$F_{t,Ed}$ $F_{t,Ed}$ $F_{c,Ed}$ $F_{c,Ed}$
10	Bolts in tension	$F_{t,Ed}$ $F_{t,Ed}$
11	Bolts in shear	$F_{v,Ed}$
12	Bolts in bearing (on beam flange, column flange, end plate or cleat)	$F_{b,Ed}$
13	Concrete in compression including grout	$F_{c,Ed}$

Table 1.2 – List of components covered by Eurocode 3 Part 1-8, (continuation)

No	Component	
14	Base plate in bending under compression	
15	Base plate in bending under tension	$F_{t,Ed}$
16	Anchor bolts in tension	$F_{t,Ed}$
17	Anchor bolts in shear	$F_{v,Ed}$
18	Anchor bolts in bearing	$F_{b,Ed}$
19	Welds	
20	Haunched beam	$F_{c,Ed}$

(a) Welded joint

(b) Bolted joint with extended end plate

(c) Two joints with extended end plates (double-sided configuration)

(d) Joint with flush end plate

(e) End-plate type beam splice

(f) Cover-joint type beam splice

(g) Bolted joint with angle flange cleats

(h) Two beam-to-beam joints (double-sided configuration)

Figure 1.20 – Examples of steel beam-to-column joints, beam-to-beam joints and beam splices joints covered in EN 1993-1-8

> **UK Specific Comment**
>
> With reference to Figure 1.20 it is noted that (site) welded joints and angle flange cleats are rarely, if ever, used in the UK.

1.6.2 Hybrid joints aspects

In the present book, the design of welded and bolted connections is mainly considered. All the principles ruling the design of such joints are globally similar for hybrid joints in which the transfer of forces between two connected elements is commonly achieved by two different connectors, for instance welds and bolts.

Hybrid joints are however not widely used in practice and many experts in the field of joints are not recommending them, because of the different level of stiffness which often characterises their respective behaviours and which generally prevents the user to "add up" the contributions of the two connection systems to the resistance and so to profit from the higher expected joints resistance. For these reasons, hybrid joints will not be explicitly addressed here.

In EN 1993-1-8, few recommendations are given: "where fasteners with different stiffnesses are used to carry a shear load, the fasteners with the highest stiffness should be designed to carry the design load. As an exception, preloaded class 8.8 and 10.9 bolts in connections designed as slip-resistant at the ULS may be assumed to share load with welds, provided that the final tightening of the bolts is carried out after the welding is complete."

> **UK Specific Comment**
>
> Hybrid connections are rarely used in the UK for the reasons noted.

1.7   DESIGN TOOLS

EN 1993 gives new and advanced options to design efficient and economic steel structures. The design of joints plays a major role in that process. Thus the detailing of joints and the methods of considering the joints properties in the frame analysis will significantly influence the costs of a steel structure. This has been demonstrated by various investigations.

However, the exploitation of the advanced possibilities is rather time consuming for the designer if no appropriate tools for a quick and easy design are available. Different opinions have been discussed in Europe concerning the development of the Eurocodes. On one side, it is expected that the Eurocodes provide design methods which will allow safe, robust and economic solutions. Of course this requires more sophisticated approaches for the design rules. On the other side, the users of the Eurocodes are requesting simple codes for practice. But this is in conflict with the major request to make steel structures more economic. It would be unfortunate to make standards too simple as there is the loss of many possibilities to take advantage of the new and advanced options mentioned above.

The message is quite clear: there was and there is still a need for sophisticated standards which form an accepted basis to design steel structures. Based on the methods given in these standards simple design tools need to be developed and provided to practitioners. This is an optimal way to bring more economic solutions to the market with an acceptable effort needed by the designers. As it is often said: *"Not simple rules sell steel, but simple tools sell steel."*

1.7.1 Types of design tools

Beside the need for background information, the engineer requires simple design tools to be able to design joint in an efficient way. Three different types of design aids can be provided. The most appropriate type depends on various aspects.

- *Design tables* are ready-to-use tables containing standardised joint layouts including dimension details and all relevant mechanical properties like resistance, stiffness and ductility. The use of tables is certainly the quickest way to design a joint. However, any change in the layout will require further calculations and tables are no more helpful;
- *Design sheets* are sets of simple design formulae. The aim is to allow a simple and rather quick hand calculation. Due to simplifications, the results could be more conservative or the range of validity more limited. Both design tables and design sheets can be published in handbooks;

– The most flexible way is the use of *software*. Of course it takes a few minutes to enter all joint details, but there will be only few limitations in the range of validity, and any re-calculation, for example due to a change in the layout, is a matter of a few seconds.

1.7.2 Examples of design tools

In recent years, efforts have been made in various countries to develop specific design tools for joints. It is not practical to be exhaustive in the referencing of these tools. Therefore only a few characteristic examples (drafted in English) will be briefly described below to illustrate the variety of the available information sources.

The difficulties for the designer are to know about the existence of such tools, and to select the one which would be the most appropriate for his daily practice. This will require some investment in time, but this investment will be quickly repaid by an efficient and economical design of the joints, a quick fabrication in workshop and an easy erection on site.

– Design tables: as an example, the DSTV/DASt Ringbuch (Weynand and Oerder, 2013) can be mentioned here. The book is a publication of Stahlbau Verlags- und Service GmbH in Germany and is published in two languages, English and German. It covers a very wide range of moment-resisting and simple (pinned) joints in various joint configurations: beam-to-column, beam splices and beam-to-beam. Bolted connections with flush and extended end plates, with two or four bolts per row, header plates and web cleats are considered, as well as different bolt and steel grades. The book consists of design tables, which can be used in a straightforward manner to select a joint or, for instance, to check the resistance of a specified joint. They are provided for a selection of standard combinations of connected member sections and, in more details, provide the designer with the following data:

  ▪ the detailing of the joint (geometric and mechanical properties of the sections and connection elements);
  ▪ indications about requirements for stiffeners;
  ▪ for simple joints: the design shear resistance of the joint;

- for moment-resisting joints: the design moment resistance under positive and negative bending moments and the shear resistance of the joint as well as the initial stiffness;
- an identification of the joint component which is governing the resistance moment;

Similar books are published in the UK as joint publications of the SCI, the British Constructional Steel Association (BCSA) and Tata Steel, for moment-resisting joints (Brown *et al*, 2013), and for simple joints (Moreno *et al*, 2011).

– Design sheets for a simple design of joints in accordance with EN 1993 was first published published by the European Commission (EC) (CRIF 1996). These sheets reference a EC-funded project, called SPRINT, in which they have been developed. Each individual set of sheet is devoted a particular joint configuration and connection type, see Figure 1.21.

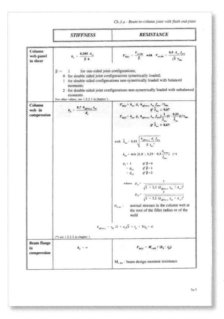

Figure 1.21 – Design sheets (CRIF 1996)

The joint design procedure included in this design tool is aimed at assisting the designer who wishes to take account of the full potential of semi-rigidity and/or partial-strength joints, without having to go through the more general but often complex approach provided by EN 1993-1-8. In reality, to derive these sheets, profit has been taken of all possible "shortcuts" allowed by the standards so as to safely but more easily perform the calculations. In each set, the first design sheet summarises all the data requested concerning the joint configuration and the connection type. In the remaining sheets, the calculation procedure first provides all the expressions for the evaluation of both stiffness and resistance for each of the joint components in a logical order, and finally shows how to derive the mechanical properties of the whole joint, i.e. the initial stiffness and/or the design moment resistance. The failure mode corresponds to the component whose resistance determines the design moment resistance of the joints. It gives an indication on the level of rotational capacity of the joint.

The shear resistance of the joint is an important value. For the sake of clarity, it is not dealt with in the design sheets but relevant information is provided just after the sheets.

In Maquoi and Chabrolin (1998), sheets are provided for beam splices and beam-to-column steel joints with end plates or flange cleats. In Anderson *et al* (1999), similar sheets are available for composite beam-to-column joints. Such sheets may be prepared for any joint configuration, connections types or loading situations. The designer may easily programme the sheet so as to develop its own set of sheets or even to establish his own design tables (i.e. corresponding to the particular need of his company).

39

– A number of software tools are offered on the market. As an example, the programme COP (Weynand *et al*, 2014), is mentioned here. COP is an innovative computer programme for the design of joints in steel and composite structures. The calculations are made in full accordance with the EN 1993-1-8 using the component method. The software fits perfectly with the needs of engineers and draughtsmen. All joint details are defined in clearly arranged data input masks. During input, all data will be visualised in scaled two-dimensional (2-D) or 3-D views and a data check module monitors if the requirements of EN 1993, such as bolt end distances, weld sizes, etc., are fulfilled, and, if needed, valid values are proposed. Four editions of COP are available for:

- Composite joints;
- Hollow section joints;
- Simple joints;
- Moment-resisting joints and simple joints.

The two first editions are freely available at http://cop.fw-ing.com. COP provides a full calculation note and it is available in English, German and French (output language independent of user interface language).

UK Specific Comment

There is not much use of CoP$^©$ in the UK. There are a number of other software tools on the market, some more comprehensive than others.

1.8 WORKED EXAMPLES

Worked examples represent another significant aid for designers as they start to apply new design approaches. However, in the field of joints, the variety of the joint configurations, connection types, connected elements and loading situations is such that a number of worked examples would be required to cover the whole range of European practical applications and design practices.

To be useful, a complete set of worked examples must cover the following:

- Bolted connections;
- Welded connections;
- Simple joints in buildings:

  - beam splice, beam-to-beam and beam-to-column joints;
  - column splices;
  - column bases;

- Moment-resisting joints in buildings:

  ▪ beam splice, beam-to-beam and beam-to-column joints;
  ▪ column splices;
  ▪ column bases;

- Joints in lattice girders;

knowing that, in each category, several sub-categories should be considered according to the connected and the connecting elements and the loading situation.

In this book, a design example of a beam-to-column joint with an end plate connection with several bolt-rows and subjected to bending and shear is given in section 6.6.3. This case illustrates the components characterisation and assembly to derive stiffness and the resistance of the whole joint.

Several publications are available for other situations which can be found in practice. Hereunder, some are selected (non-exhaustive list) according to the two following criteria: easily available to designers and drafted in English.

- Bolted and welded connections (Veljkovic, 2015);
- Simple joints in buildings: beam splice, beam-to-beam and beam-to-column joints (Anderson *et al*, 1999; Moreno *et al*, 2011; Veljkovic, 2015);
- Simple joints in buildings: column splices (Brettle, 2009; Moreno *et al*, 2011);

41

- Simple joints in buildings: column bases (Brettle, 2009; Moreno *et al*, 2011; Veljkovic, 2015);
- Moment-resisting joints in buildings: beam splice, beam-to-beam and beam-to-column joints (Anderson *et al*, 1999; Brettle, 2009; Brown *et al*, 2013; Veljkovic, 2015);
- Moment-resisting joints in buildings: column splices (Brettle, 2009; Brown *et al*, 2013);
- Moment-resisting joints in buildings: column bases (Brettle, 2009; Brown *et al*, 2013; Veljkovic, 2015);
- Joints in lattice girders (Wardenier *et al*, 1991; Packer *et al*, 1992; Kurobane *et al*, 2004; Wardenier *et al*, 2008; Brettle, 2008; Packer *et al*, 2009).

Chapter 2

STRUCTURAL ANALYSIS AND DESIGN

2.1   INTRODUCTION

The design of a frame and its components is a two-step procedure involving a global frame analysis followed by individual cross section and member design checks.

Global frame analysis is aimed at deriving the values of the internal forces and moments and of the displacements in the given structure, when subjected to a given set of loads. It is based on assumptions regarding the component behaviour (elastic or plastic) and the geometric response (first-order or second-order theory) of the frame. The ECCS Eurocode Design Manual for the Design of Steel Structures (Simões da Silva *et al*, 2010) provides detailed information about the different available analysis procedures. Once the analysis is complete, i.e. all relevant internal forces and moments and displacements are determined in the whole structure, then the design checks of all the frame components can be performed. These ones consist in verifying whether the structure satisfies all the required design criteria under service loads (SLS) and under factored loads (ULS).

Globally speaking, four main analysis approaches may be considered according to EN 1993 (see Table 2.1):

- Linear elastic first order analysis;
- Elastoplastic first order analysis;
- Linear elastic second order analysis;
- Elastoplastic second order analysis;

Table 2.1 – Four main approaches for frame analysis

		Assumed material law (linear or non-linear)	
		Elastic	Plastic
Assumption in terms of geometric effects (linearity or non-linearity)	1st order	Linear elastic first order analysis	Elastoplastic first order analysis
	2nd order	Linear elastic second order analysis	Elastoplastic second order analysis

The selection of the most appropriate methodology is quite important, and will strongly influence the number and type of checks to be carried out, as already stated in section 1.1.1. It should be noted however that the designer may not have complete freedom of choice. The following paragraphs summarise the recommendations given in EN 1993 in this domain.

2.1.1 Elastic or elastoplastic analysis and verification process

The selection of an elastic or plastic analysis and design process depends on the class of the member cross sections (Class 1, 2, 3 or 4). A class expresses the way on how the possible local plate buckling of cross section walls subjected to compression may or not affect the resistance or the ductility of the cross sections.

An elastic analysis is required for frames made of members with Class 2, Class 3 or Class 4 cross sections. For frames with Class 1 member cross sections (at least at the locations of the plastic hinges), a elastoplastic analysis may be also performed.

The cross section and member design checks are also depending on the cross section class. Table 2.2 summarises the various possibilities given in EN 1993.

Table 2.2 – Selection between elastic and elastoplastic analysis and design process

Cross section class	Frame analysis	Cross section/member verifications	Designation of the global approach (E = elastic; P = plastic)
Class 1	Plastic	Plastic	P-P
	Elastic	Plastic	E-P
	Elastic	Elastic	E-E
Class 2	Elastic	Plastic	E-P
	Elastic	Elastic	E-E
Class 3 – Class 4	Elastic	Elastic	E-E

2.1.2 First order or second order analysis

Geometric non-linearities may affect the response of members or the response of the full frame and lead respectively to local (flexural buckling, torsional buckling, lateral torsional buckling) or global sway instability phenomena.

Structural analysis considers the external forces acting on the structure and the resulting internal forces and moments in the members needed to maintain equilibrium. According to the geometry of the structure that is used to express this equilibrium, different analysis options are available:

- In a first order analysis, reference is made to the initial undeformed shape of the structure;
- In a second order analysis, reference is made to the actual deformed shape of the structure.

Two types of second order behaviour have to be considered, linked to so-called $P - \Delta$ effects and $P - \delta$ effects. $P - \Delta$ effects determine possible sway instability phenomena whereas $P - \delta$ effects govern the possible development of member instabilities. A full second order analysis will be characterised by an explicit consideration of all geometric second order effects (sway and member). In this case, no member check has to be performed further to the structural analysis. However, in most cases $P - \delta$ effects are integrated in the analysis to cover buckling phenomena but not the lateral torsional buckling of the members. So, an additional check for this potential instability mode may still be needed.

Usually, the designer prefers not to integrate the so-called $P - \delta$ effects (second order effects related to the member buckling) in the frame analysis and so to check member buckling phenomena further to the analysis, when verifying ULS. It just then remains to decide whether the $P - \Delta$ effects are to be or not considered by means of a specific criterion defining the "sway" or "non sway" character of the structure ("α_{cr}" criterion in EN 1993-1-1).

The description "non-sway frame" applies to a frame when its response to in-plane horizontal forces is so stiff that it is acceptable to neglect any additional forces or moments arising from horizontal displacements of its storeys. This means that the global second-order effects may be neglected. When the second order effects are not negligible, the frame is said to be a "sway frame".

Figure 2.1 summarises the choices offered to the designer, not only in terms of frame analysis, but also of type of cross section and/or member verification at ULS further to the frame analysis.

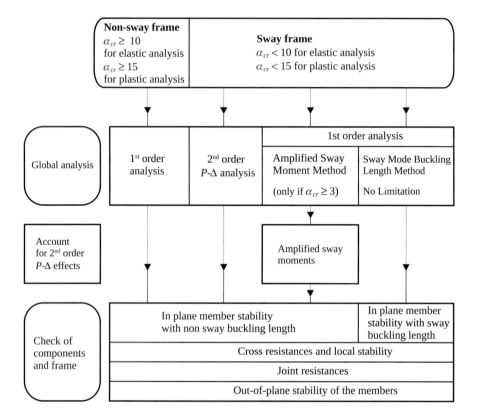

Figure 2.1 – Various ways for the global analysis and design process

2.1.3 Integration of joint response into the frame analysis and design process

Traditionally, frame analysis is performed by assuming joints as perfectly pinned or perfectly rigid. It is important that the joints are subsequently designed and detailed to behave in a way that corresponds to the analysis assumptions. In chapter 1 it has been pointed out that significant global economy may often result from what has been presented as a semi-continuous approach for structural joints.

A cross section being nothing else than a particular cross section, characterised by its own stiffness, resistance and ductility, the integration of the actual joint properties into the frame analysis and design process appears as natural. And it is the case. For instance, all what has been said in the previous sections remains unchanged at the condition that the meaning of the word "cross section" is extended to "member or joint cross section". From an operational point of view, the attention has however to be drawn to some specific concepts respectively named "joint modelling", "joint classification" and "joint idealisation" that are presented in the following sections.

2.2   JOINT MODELLING

2.2.1 General

Joint behaviour affects the structural frame response and, therefore, has to be modelled, just like beams and columns are modelled for frame analysis and design. Traditionally, the following types of *joint modelling* are considered:

– *For rotational stiffness:*

- rigid
- pinned

– *For* resistance*:*

- full-strength
- partial-strength
- pinned

When the joint rotational stiffness is being considered, the wording *rigid* means that no relative rotation occurs between the connected members whatever the applied moment. The wording *pinned* postulates the existence of a perfect (i.e. frictionless) hinge between the members. In fact, these definitions may be relaxed, as explained in sub-chapter 2.4 devoted to the joint classification. Indeed, rather flexible but not fully pinned joints, and rather stiff but not fully rigid joints may be considered as fairly pinned and fairly rigid, respectively. The stiffness boundaries for joints to be classified as rigid or pinned are defined in sub-chapter 2.4.

In terms of the joint resistance, a *full-strength joint* is stronger than the weaker of the connected members, as opposed to a *partial-strength joint*. In the daily practice, partial-strength joints are used whenever the joints are designed to transfer the internal forces and not to resist the full capacity of the connected members. A *pinned joint* cannot transfer moment. Related classification criteria are conceptually discussed in sub-chapter 2.4.

Consideration of rotational stiffness and resistance joint properties leads to three significant types of joint modelling:

- Rigid/full-strength;
- Rigid/partial-strength;
- Pinned.

However, in terms of joint rotational stiffness, joints designed for economy may be neither rigid nor pinned but semi-rigid. There are thus new possibilities for joint modelling:

- Semi-rigid/full-strength;
- Semi-rigid/partial-strength.

For simplification, EN 1993-1-8 accounts for these possibilities by introducing three joint models (Table2.3):

- *Continuous* covering the rigid/full-strength case only;
- *Semi-continuous* covering the rigid/partial-strength, the semi-rigid/full-strength and the semi-rigid/partial-strength cases;
- *Simple* covering the pinned case only.

Table 2.3 – Types of joint modelling

Stiffness	Resistance		
	Full-strength	Partial-strength	Pinned
Rigid	Continuous	Semi-continuous	*
Semi-rigid	Semi-continuous	Semi-continuous	*
Pinned	*	*	Simple
	* : No meaning		

The following meanings are given to these terms:

- *Continuous:* the joint ensures a full rotational continuity between the connected members;
- *Semi-continuous:* the joint ensures only a partial rotational continuity between the connected members;
- *Simple*: the joint prevents from any rotational continuity between the connected members.

The interpretation to be given to these wordings depends on the type of frame analysis to be performed. In the case of an elastic global frame analysis, only the stiffness properties of the joint are relevant for the joint modelling. In the case of a rigid-plastic analysis, the main joint feature is the resistance. In all the other cases, both the stiffness and resistance properties govern the way the joints are modelled. These possibilities are illustrated in Table 2.4.

Table 2.4 – Joint modelling and frame analysis

Modelling	Type of frame analysis		
	Elastic analysis	Rigid-plastic analysis	Elastic-perfectly plastic and elastoplastic analysis
Continuous	Rigid	Full-strength	Rigid/full-strength
Semi-continuous	Semi-rigid	Partial-strength	Rigid/partial-strength Semi-rigid/full-strength Semi-rigid/partial-strength
Simple	Pinned	Pinned	Pinned

2.2.2 Modelling and sources of joint deformation

As an example, the difference between the behaviour under load of the connection and that of the column web in a beam-to-column joint (see section 1.2.2) requires, from a theoretical point of view, that account be taken separately of both deformation sources when designing a building frame. However, doing so is only feasible when the frame is analysed by means of a sophisticated computer programme which enables a separate modelling of both deformation sources. For most available software, the modelling of the joints has to be simplified by concentrating the sources of deformation into a single rotational spring located at the intersection of the axes of the connected members.

2.2.3 Simplified modelling according to EN 1993

For most applications, modelling the connection separately from the web panel behaviour is neither useful nor feasible; therefore only the simplified modelling of the joint behaviour (see section 2.2.2) will be considered in EN 1993-1-8. Table 2.5, excerpted from EN 1993-1-8, shows how to relate the simplified modelling of typical joints to the basic terminology used for the joint modelling: simple, semi-continuous and continuous.

Table 2.5 – Simplified modelling for joints according to EN 1993

Joint modelling	Beam-to-column joints major axis bending	Beam splices	Column bases
Simple			
Semi-continuous			
Continuous			

2.2.4 Concentration of the joint deformation

It is not practical to separate the flexural behaviour of the connection and the shear (major axis beam-to-column joint) or out-of-plane (minor axis beam-to-column joint configurations or beam-to-beam configurations) behaviour of the column web panel. This section explains how to concentrate the two deformabilities into a single flexural spring located at the intersection of the axes of the connected members.

2.2.4.1 Major axis beam-to-column joint configurations

In a single-sided configuration there is only one joint to be considered. The characteristic deformation curve of the column web panel in shear (see Figure 1.9 and Figure 2.2b) is first transformed into a $M_b - \gamma$ curve by means of the *transformation parameter β*. This parameter is defined in Figure 2.3a and relates the web panel shear force to the (load-introduction) compressive and tensile connection forces, see Eqs. (1.1) and (1.2).

The $M_b - \phi$ spring characteristic which represents the joint behaviour is shown in Figure 2.2c; it is obtained by summing the contributions of rotation, from the connection, ϕ_c, and from the shear panel, γ. The $M_j - \phi$ characteristic of the joint rotational spring located at the beam-to-column interaction is assumed to be equal to the $M_j - \phi$ characteristic obtained as indicated in Figure 2.2c.

51

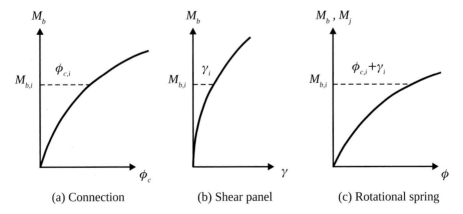

| (a) Connection | (b) Shear panel | (c) Rotational spring |

Figure 2.2 – Flexural characteristic of the rotational spring

$V_{wp} = \beta F_b$
where $F_b = M_b /z$

$V_{wp} = \beta_1 F_{b1}$
$= \beta_2 F_{b2}$
where $F_{b1} = M_{b1}/z$
$F_{b2} = M_{b2}/z$

(a) Single-sided joint configuration   (b) Double-sided joint configuration

Figure 2.3 – Definition of the transformation parameter β

In a double-sided configuration there are two joints to be considered, the left joint and the right joint. The derivation of their corresponding deformation curves is similar to the single-sided configurations by using transformation parameters β_1 and β_2 (Figure 2.3b).

As already said in section 1.2.2, significant shear forces are likely to develop in the column web panels. They result from all the forces acting in the adjacent beams and columns, i.e. in a double-sided joint configuration (see Figure 2.4) the shear force in the web panel $V_{wp,Ed}$ is determined as:

$$V_{wp,Ed} = \frac{\left(M_{b1,Ed} - M_{b2,Ed}\right)}{z} - \frac{\left(V_{c1,Ed} - V_{c2,Ed}\right)}{2} \tag{2.1}$$

where z designates the level arm of the internal forces (see Figure 2.5).

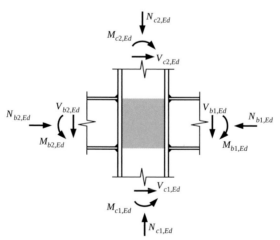

Figure 2.4 – Forces applied at the periphery of a web panel

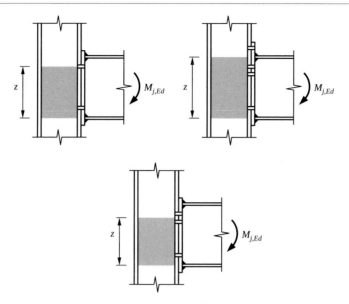

Figure 2.5 – Definition of the level arm z

Figure 2.6 illustrates the determination of the shear force in a column web panel for few particular cases.

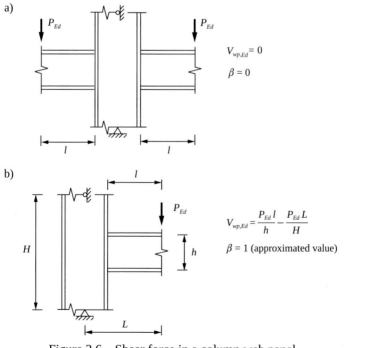

a)

$V_{wp,Ed} = 0$

$\beta = 0$

b)

$V_{wp,Ed} = \dfrac{P_{Ed}\, l}{h} - \dfrac{P_{Ed}\, L}{H}$

$\beta = 1$ (approximated value)

Figure 2.6 – Shear force in a column web panel

53

c)

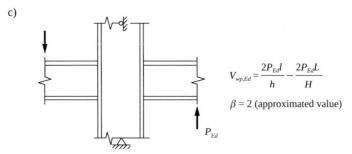

$$V_{wp,Ed} = \frac{2P_{Ed}l}{h} - \frac{2P_{Ed}L}{H}$$

$\beta = 2$ (approximated value)

P_{Ed}

Figure 2.6 – Shear force in a column web panel (continuation)

The β coefficients are defined as (see Figure 2.3b):

$$\beta_1 = \frac{V_{wp,Ed} z}{M_{b1,Ed}} \quad \text{for the right joint} \tag{2.2}$$

$$\beta_2 = \frac{V_{wp,Ed} z}{M_{b2,Ed}} \quad \text{for the left joint} \tag{2.3}$$

Because the values of the β parameter can only be determined once the internal forces are known, their accurate determination requires an iterative process in the global analysis. For practical applications, such an iterative process may be avoided provided that conservative β values are available. These values may be used a priori to model the joints and, on the basis of such joint modelling, the frame analysis may be performed safely in a non-iterative way.

The recommendations for the approximate values of β are given in Table 5.4 of EN 1993-1-8, see Table 2.6. They vary from $\beta = 0$ (double-sided joint configuration with balanced moments in the beams) to $\beta = 2$ (double-sided joint configuration with equal but unbalanced moments in the beams). These two extreme cases are illustrated in Figure 2.7.

Table 2.6 – Approximate values for the transformation parameter β

Type of joint configuration		Action	Value
$M_{b1,Ed}$ $M_{b1,Ed}$		$M_{b1,Ed}$	$\beta \approx 1$

Table 2.6 – Approximate values for the transformation parameter β (continuation)

	$M_{b1,Ed} = M_{b2,Ed}$	$\beta = 0^{*)}$
	$M_{b1,Ed}/M_{b2,Ed} > 0$	$\beta \approx 1$
	$M_{b1,Ed}/M_{b2,Ed} < 0$	$\beta \approx 2$
	$M_{b1,Ed} + M_{b2,Ed} = 0$	$\beta \approx 2$

[*)] In this case the value of β is the exact value rather than an approximation

(a) Balanced beam moments   (b) Equal but unbalanced beam moments

Figure 2.7 – Extreme cases for β values

The values given in Table 2.6 are derived from the following simplified version of Eqs. (2.2) and (2.3):

$$\beta_1 = \left|1 - M_{b2,Ed}/M_{b1,Ed}\right| \leq 2 \quad \text{for the right joint} \tag{2.4}$$

$$\beta_2 = \left|1 - M_{b1,Ed}/M_{b2,Ed}\right| \leq 2 \quad \text{for the left joint} \tag{2.5}$$

In these formulations, the contribution of the shear forces acting in the column is conservatively disregarded (Jaspart, 1991).

2.2.4.2 Minor axis beam-to-column joint configurations and beam-to-beam configurations

Similar concepts to those developed in section 2.2.4.1 are referred to for minor axis beam-to-column joint configurations and beam-to-beam configurations. The definition of the transformation parameter is somewhat different (see Figure 2.8).

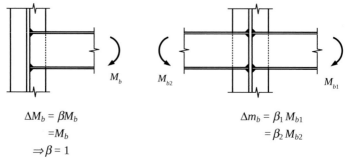

$$\Delta M_b = \beta M_b$$
$$= M_b$$
$$\Rightarrow \beta = 1$$

(a) Single-sided joint configuration

$$\Delta m_b = \beta_1 M_{b1}$$
$$= \beta_2 M_{b2}$$

(a) Double-sided joint configuration

Figure 2.8 – Definition of the transformation parameter β

UK Specific Comment

Although in UK design practice most aspects related to joint modelling have developed to a high degree of sophistication, proper consideration of the structural interaction between the joints and members is rarely fully taken into account. It is indeed very difficult to predict and model the complex behaviour of a joint (strength, stiffness and rotation capacity). Traditional UK practice is to predict joint behaviour on the basis of past experience.

BSI (2010) separates the modelling of the rotational behaviour of the connection elements (that is, the bolts, the welds, the plates, etc.) and that of the column web panel in shear to characterise the overall joint behaviour. As a simplification, the two sources of joint deformation are combined by means of the so-called transformation parameter β This coefficient is related to the web panel internal actions and can be calculated from the moments at each side of the web panel. The β parameter is currently only defined for same-depth beams which meet at the same level. In many practical double-sided joints, beams will not be of the same depth, or at the same height. BS EN 1993-1-8 is silent on such joints.

2.3 JOINT IDEALISATION

The non-linear behaviour of the isolated flexural spring which characterises the actual joint response does not lend itself towards everyday design practice. However the $M - \phi$ characteristic curve may be idealised without significant loss of accuracy. One of the most simple idealisations possible is the elastic-perfectly-plastic one (Figure 2.9a). This modelling has the advantage of being quite similar to that used traditionally for the modelling of member cross sections subject to bending (Figure 2.9b).

The moment $M_{j,Rd}$ that corresponds to the yield plateau is termed design moment resistance in EN 1993. It may be understood as the *pseudo-plastic moment resistance* of the joint. Strain-hardening effects and possible membrane effects are thus neglected; that explains the difference in Figure 2.9 between the actual $M - \phi$ characteristic and the *yield plateau* of the idealised one.

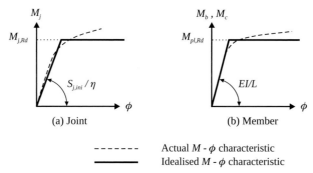

(a) Joint   (b) Member

- - - - - - -   Actual $M - \phi$ characteristic
————————   Idealised $M - \phi$ characteristic

Figure 2.9 – Bi-linearisation of moment-rotation curves

The value of the constant stiffness is discussed below. In fact there are different possible ways to idealise a joint $M - \phi$ characteristic. The choice of one of them depends on the type of frame analysis that is considered, as explained below.

2.3.1 Elastic idealisation for an elastic analysis

The main joint characteristic is the constant rotational stiffness.

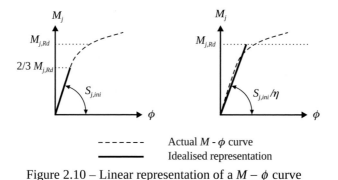

- - - - - - -   Actual $M - \phi$ curve
————————   Idealised representation

Figure 2.10 – Linear representation of a $M - \phi$ curve

Two possibilities are offered in EN 1993-1-8, see Figure 2.10:

- *Elastic verification of the joint resistance* (Figure 2.10a): the constant stiffness is taken equal to the initial stiffness $S_{j,ini}$; at the end of the frame analysis, it has to be checked that the design moment $M_{j,Ed}$ experienced by the joint is less than the maximum elastic joint moment resistance defined as $2/3M_{j,Rd}$;
- *Plastic verification of the joint resistance* (Figure 2.10b): the constant stiffness is taken equal to a fictitious stiffness, the value of which is intermediate between the initial stiffness and the secant stiffness relative to $M_{j,Rd}$; it is defined as $S_{j,ini}/\eta$. This idealisation is aimed at "replacing" the actual non-linear response of the joint by an "equivalent" constant one; it is valid for $M_{j,Ed}$ values less than or equal to $M_{j,Rd}$. EN 1993-1-8 recommends values of η as in Table 2.7.

Table 2.7 – Stiffness modification factor η

Type of connection	Beam-to-column joints	Other types of joints (beam-to-beam joints, beam splices, column base joints)
Welded	2	3
Bolted end-plates	2	3
Bolted flange cleats	2	3.5
Base plates	-	3

2.3.2 Rigid-plastic idealisation for a rigid-plastic analysis

Only the design resistance $M_{j,Rd}$ is required for this type of analysis. In order to allow the possible plastic hinges to form and rotate in the joint locations, it has to be checked that the joint has a sufficient rotation capacity, see Figure 2.11.

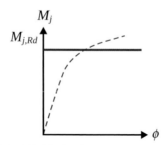

Figure 2.11 – Rigid-plastic representation of a $M - \phi$ curve

2.3.3 Non-linear idealisation for an elastic-plastic analysis

The stiffness and resistance properties are equally important in this case. The possible idealisations range from bi-linear, tri-linear representations or a fully non-linear curve, see Figure 2.12. Again, rotation capacity is required in joints where plastic hinges are likely to form and rotate.

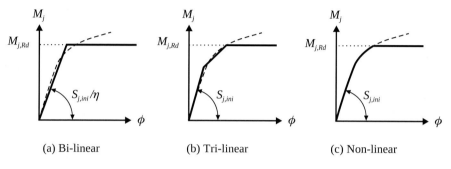

(a) Bi-linear	(b) Tri-linear	(c) Non-linear

Figure 2.12 – Non-linear representations of a $M - \phi$ curve

---

UK Specific Comment

As explained above, typical UK design practice does not consider this sophisticated procedure for idealisation of joints.

2.4 JOINT CLASSIFICATION

2.4.1 General

In sub-chapter 2.2, it is shown that the joints need to be modelled for the global frame analysis and that three different types of joint modelling are introduced: simple, semi-continuous and continuous. It has also been explained that the nature and details of joint modelling are dependent both on the type of frame analysis and on the class of the joint in terms of stiffness and/or strength (Table 2.4).

Classification criteria are used to define the stiffness class and the strength class to which the joint belongs and also to determine the type of joint modelling which shall be adopted for analysis. They are described below.

59

2.4.2 Classification based on mechanical joint properties

The joint is classified in terms of stiffness by comparing the design value joint stiffness with two stiffness boundaries (Figure 2.13). For the sake of simplicity, the stiffness boundaries have been derived so as to allow a direct comparison with the initial design joint stiffness, whatever the type of joint idealisation that is used afterwards in the analysis (Figures 2.10 and 2.12).

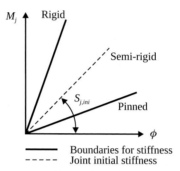

Figure 2.13 – Stiffness classification boundaries

The strength classification simply consists in comparing the joint design moment resistance to full-strength and pinned boundaries (Figure 2.14).

Figure 2.14 – Strength classification boundaries

The stiffness and strength boundaries for the joint classification are given in Table 2.8. The classification as rigid or pinned of joints, exhibiting an actual stiffness not exactly equal to ∞ or 0, is based on the recognition that the actual structural response, in terms of load-carrying capacity, will never differ from more than 5% from the one evaluated by the designer on the basis of the structure with perfectly rigid or pinned joints.

Table 2.8 – Joint classification boundaries

Classification by stiffness

Beam-to-column joint in an unbraced frame[**]

    Pinned joint: $S_{j,ini} \leq 0.5\ EI_b/L_b$

    Semi-rigid joint: $0.5\ EI_b/L_b < S_{j,ini} \leq 25\ EI_b/L_b$

    Rigid joint: $S_{j,ini} \geq 25\ EI_b/L_b$

Beam-to-column joint in a braced frame[***] :

    Pinned joint: $S_{j,ini} \leq 0.5\ EI_b/L_b$

    Semi-rigid joint: $0.5\ EI_b/L_b < S_{j,ini} \leq 8\ EI_b/L_b$

    Rigid joint: $S_{j,ini} \geq 8\ EI_b/L_b$

Beam splices[****] :

    Pinned joint: $S_{j,ini} \leq 0.5\ EI_b/L_b$

    Semi-rigid joint: $0.5\ EI_b/L_b < S_{j,ini} \leq 25\ EI_b/L_b$

    Rigid joint: $S_{j,ini} \geq 25\ EI_b/L_b$

Column bases in non-sway frames[***] :

    Rigid joint: if $S_{j,ini} \geq 0$

                   if $0.5 < \overline{\lambda}_0 < 3.93$ $S_{j,ini} \geq 7\left(2\overline{\lambda}_0 - 1\right)EI_c/L_c$

                   if $\overline{\lambda}_0 \geq 3.93$ $S_{j,ini} \geq 48\ EI_c/L_c$

Column bases in sway frames[**] :

    Rigid joint : $S_{j,ini} \geq 30\ EI_c/L_c$

Important remark: the validity of the stiffness classification for beam-to-column joints is restricted to structures where possible beam splices are classified as rigid. Fortunately, this limitation is not very restrictive for practical applications.

[**]   Frames not fulfilling [***], provided that in every storey $K_b/K_c \geq 0.1$ where:
      K_b is the mean value of I_b/L_b for all the beams at the top of that storey;
      K_c is the mean value of I_c/L_c for all the columns in that storey.
      For frames where $K_b/K_c < 0.1$, the joints should be classified as semi-rigid.

[***]   Frame where the bracing system reduces the horizontal displacements by at least 80%.

[****] Limit not specified in EN 1993-1-8.

Table 2.8 – Joint classification boundaries (continuation)

Classification by resistance	
Full strength joint:	Joint which possesses a higher resistance than the weakest of the connected members.
Pinned joint:	Joint with design resistance lower than 25 % of the full-strength resistance.
Partial-strength joint:	Joint with design resistance lower than the full-strength resistance and which has not been classified as pinned.

UK Specific Comment

The NA to BS EN1993-1-8 gives practical guidance to classify joints. It also states the following:

"Until experience is gained with the numerical method of calculating rotational stiffness given in BS EN 1993-1-8:2005, 6.3 and the classification by stiffness method given in BS EN 1993-1-8:2005, 5.2.2, semi-continuous elastic design should only be used where either it is supported by test evidence according to BS EN 1993-1-8:2005, 5.2.2.1(2) or where it is based on satisfactorily performance in a similar situation."

Evidence on values of column base stiffness for practical global analysis is given in SCI (2006).

62

2.5   DUCTILITY CLASSES

2.5.1 General concept

Experience and proper detailing result in so-called *pinned* joints which exhibit a sufficient rotation capacity to sustain the rotations imposed on them. This topic is addressed in chapter 5.

For moment-resisting joints the concept of ductility classes is introduced to deal with the question of rotation capacity.

For most of these structural joints, the shape of the $M - \phi$ characteristic can be approximated by a bi-linear relationship (Figure 2.15a). The initial slope $S_{j,ini}$ corresponds to the elastic deformation of the joint. It is followed by progressive yielding of the joint until the design moment resistance $M_{j,Rd}$ is reached. Then a post-limit behaviour ($S_{j,post-lim}$) develops which corresponds to the onset of strain-hardening and possibly of membrane effects. These are especially important in components where rather thin plates are subject to transverse tensile forces as, for instance, in minor axis joints and in joints with columns made of rectangular hollow sections.

In many experimental tests (Figure 2.15a) the collapse of the joints at a peak moment $M_{j,u}$ has practically never been reached because of high local deformations in the joints involving extremely high relative rotations. In other cases (Figure 2.15b) the collapse has involved excessive yielding (rupture of the material) or, more often, the instability of one of the constituent components (e.g. column web panel in compression or buckling of the beam flange and web in compression) or the brittle failure in the welds or in the bolts.

In some joints, the premature collapse of one of the components prevents the development of a high moment resistance and high rotation. The post-limit range is rather limited and the bi-linear character of the $M_j - \phi$ response is less obvious to detect (Figure 2.15c).

As explained in sub-chapter 2.3, the actual $M_j - \phi$ curves are idealised before performing the global analysis. As for beam and column cross sections, the usual concept of plastic hinge can be referred to for plastic global analysis.

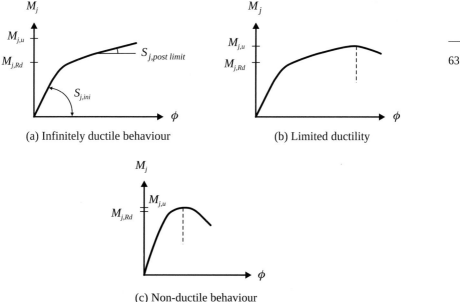

(a) Infinitely ductile behaviour   (b) Limited ductility

(c) Non-ductile behaviour

Figure 2.15 – Shape of joint $M_j - \phi$ characteristics

The development of plastic hinges during the loading of the frame and the corresponding redistribution of internal forces in the frame require, from the joints where hinges are likely to occur, a sufficient rotation capacity. In other words, there must be a sufficiently long yield plateau ϕ_{pl} (Figure 2.16) to allow the redistribution of internal forces to take place.

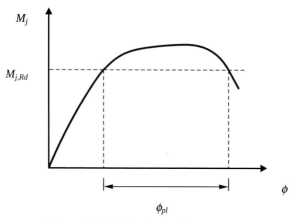

Figure 2.16 – Plastic rotation capacity

For beam and column sections, deemed-to-satisfy criteria allow the designer to determine the class of the sections and therefore the type of global frame analysis which can be considered. A strong similarity exists for what regards structural joints and a similar classification can be proposed:

- *Class 1 joints*: $M_{j,Rd}$ is reached by full plastic redistribution of the internal forces within the joints and a sufficiently good rotation capacity is available to allow, without specific restrictions, a plastic frame analysis and design to be performed, if required;
- *Class 2 joints*: $M_{j,Rd}$ is reached by full plastic redistribution of the internal forces within the joints but the rotation capacity is limited. An elastic frame analysis, possibly combined with a plastic verification of the joints, has to be performed. A plastic frame analysis is also allowed as long as it does not result in a very high required rotation capacity in the joints where hinges are likely to occur. The available and required rotation capacities have therefore to be compared before validating the analysis;
- *Class 3 joints*: brittle failure (or instability) limits the moment resistance and does not allow a full redistribution of the internal

forces in the joints. It is compulsory to perform an elastic verification of the joints.

As the moment design resistance $M_{j,Rd}$ is known whatever the collapse mode and the resistance level, no Class 4 has to be defined as it is done for member sections.

2.5.2 Requirements for classes of joints

In EN 1993 the procedure given for the evaluation of the design moment resistance of any joint provides the designer with other information such as:

- the collapse mode;
- the internal forces in the joint at collapse.

The designer knows directly whether the full plastic redistribution of the forces within the joint has been reached – the joint is then Class 1 or 2 – or not – the joint is then classified as Class 3.

For Class 1 or 2 joints, knowing the collapse mode, and more especially the component leading to collapse, gives an indication about whether there is adequate rotation capacity for a global plastic analysis to be permitted. The related criteria are detailed in chapter 6.

UK Specific Comment

The concept of ductility classes is not widely applied in the UK. It is generally assumed that standard connection details to the SCI Green Books provide enough rotation capacity and so behave in a ductile manner, as evidenced by physical testing and experience.

Chapter 3

CONNECTIONS WITH MECHANICAL FASTENERS

3.1   MECHANICAL FASTENERS

The basic elements of joints in steel structures are mechanical fasteners like bolts or pins. Chapter 3 of EN 1993-1-8 provides design rules for such mechanical fasteners: bolts, injections bolts, anchor bolts, rivets and pins. Beside these mechanical fasteners, other fasteners for more special applications can be used directly covered in EN 1993, for example flow drill bolts, HRC bolts (also called TC bolts), or nails. However, the current version of EN 1993 does not provide application rules for those types of mechanical fasteners. In some cases, harmonised product standards may provide relevant data.

In steel construction, the most typical mechanical fasteners to connect plates or profiles are bolts, or more precisely: bolt assemblies (sets) including the bolt itself, a nut and one or more washers, see Figure 3.1.

Figure 3.1 – Bolt assemblies

The bolts may be preloaded to improve serviceability performance or fatigue resistance. Joints made with preloaded bolts normally have a slight higher stiffness, but this effect is not taken into account in the design rules. However, preloading requires a controlled tightening which leads to additional work during erection.

All bolts, nut and washers should comply with the standards listed in section 1.2.4 of EN 1993-1-8. Most important standards are EN 15048 (Non-preloaded structural bolting assemblies) for non-preloaded bolts (CEN, 2007b) and EN 14399 (High strength structural bolting for preloading) for preloaded bolts (CEN, 2015). Mechanical properties are specified in EN ISO 898 (CEN, 2013). The design rules in EN 1993-1-8 are valid for all bolt classes listed in Table 3.1, but the National Annex may exclude certain bolt classes.

Table 3.1 – Bolts classes and nominal values of yield strength f_{yb} and ultimate tensile strength f_{ub}

Bolt class	4.6	4.8	5.6	5.8	6.8	8.8	10.9
f_{yb} (N/mm^2)	240	320	300	400	480	640	900
f_{ub} (N/mm^2)	400	400	500	500	600	800	1000

The name of the bolt classes indicates the nominal values of the ultimate tensile strength f_{ub} and the yield strength f_{yb} as follows. The name consists of two numbers separated by a dot. The first number is the ultimate tensile strength f_{ub} divided by 100. The fractional part indicates the ratio f_{yb}/f_{ub}, for example for a 10.9 bolt, the ultimate tensile strength f_{ub} is $10 \times 100 = 1000$ (N/mm^2) and the yield strength f_{yb} is obtained as $1000 \times 0.9 = 900$ (N/mm^2).

For preloaded bolts, only bolt assemblies of classes 8.8 and 10.9 may be used. Requirements for controlled tightening are given in EN 1090-2.

Bolt areas for common sizes of structural bolts are given in Table 3.2, where A is the gross section area of the bolt and A_s is the tensile stress area (threaded portion) of the bolt.

Table 3.2 – Bolt areas in accordance with EN ISO 898 (CEN, 2013)

d (mm)	10	12	14	16	18	20	22	24	27	30	36
A (mm^2)	78	113	154	201	254	314	380	452	573	707	1018
A_s (mm^2)	58	84	115	157	192	245	303	353	459	561	817

UK Specific Comment

The NA to BS EN 1993-1-8 excludes bolts of classes 5.8 and 6.8. It also states that bolt class 4.8 may only be used if the manufacturing process (i) is in accordance with BS EN ISO 898-1 (BSI, 2013), and (ii) includes suitable procedures to avoid hydrogen embrittlement.

3.2   CATEGORIES OF CONNECTIONS

The very basic connection is composed of a bolt assembly connecting two (or more) steel plates as shown in Figure 3.2. For the design of these basic components (for example bolts in shear and/or tension, plate in bearing), see sub-chapter 3.4.

Figure 3.2 – Bolted connection

EN 1993-1-8 distinguishes different categories of bolted connections. The two main categories, dependent on the loading applied to the bolt, are shear connections and tension connections.

3.2.1 Shear connections

Shear connections (the bolt is subjected to shear) are subdivided into three sub-categories:

a)   Category A: Bearing type

In a bearing type connection, the applied load $F_{v,Ed}$ (design shear force per bolt for the ULS) will be transferred from the plate to the bolt (shank and/or threaded part) by bearing, while the bolt will be loaded in shear. Hence, the design resistance of a category A connection is the minimum of the design bearing resistance of the plate(s) $F_{v,Rd}$ and the design shear resistance of the bolt $F_{b,Rd}$, i.e. the following criteria should be checked:

$$F_{v,Ed} \leq F_{v,Rd}$$
$$F_{v,Ed} \leq F_{b,Rd}$$

(3.1)

All bolt classes from Table 3.1 may be used. Preloading or any special precision for the contact surfaces is not required. Depending on the number of plates, the shear load applied to the bolt may be divided into one or more shear planes.

b)   Category B: Slip-resistant at serviceability limit state

With regards to the design resistance at ultimate limit state, a category B connection should meet the same requirement as a category A connection. In addition, at SLS, the connection should be designed so that no slip will occur. Therefore, preloaded bolts should be used. The design criteria are as follow:

$$F_{v,Ed} \quad \leq F_{v,Rd}$$
$$F_{v,Ed} \quad \leq F_{b,Rd}$$
$$F_{v,Ed,ser} \leq F_{s,Rd,ser}$$

(3.2)

where:
   $F_{v,Ed,ser}$ is the design shear force per bolt for SLS
   $F_{s,Rd,serv}$ is the design slip resistance per bolt at SLS.

c)   Category C: Slip-resistant at ultimate limit state

In this category, slip should not occur at ULS. This will result in the most severe design criteria for a shear connection. The design ultimate shear load will be transferred by friction between the connected plates only. This means, in theory, the connection only needs to be checked for slip resistance. In addition, EN 1993-1-8 requires checking the bearing resistance. So, the following design criteria should be checked:

$$F_{v,Ed} \leq F_{s,Rd}$$
$$F_{v,Ed} \leq F_{b,Rd} \tag{3.3}$$

where:

$F_{s,Rd}$ is the design slip resistance per bolt at SLS.

For category A and category B connections, if the connected plates are loaded in tension, the design tension resistance $N_{t,Rd}$ should be checked at ULS as well:

$$F_{v,Ed} \leq N_{t,Rd} \tag{3.4}$$

For category C connections, a similar check has also to be achieved, but on the basis of the design plastic resistance of the net cross section at the bolt holes $N_{net,Rd}$:

$$F_{v,Ed} \leq N_{net,Rd} \tag{3.5}$$

where:

$N_{t,Rd}$ is the smaller of the design plastic resistance of the gross cross section $N_{pl,Rd}$ and the design ultimate resistance of the net cross section at holes for fasteners $N_{u,Rd}$

$N_{net,Rd}$ is the design tension resistance of the net section at holes for fasteners

with

$$N_{pl,Rd} = \frac{A\,f_y}{\gamma_{M0}}, N_{u,Rd} = \frac{0.9A_{net}\,f_u}{\gamma_{M2}} \text{ and } N_{net,Rd} = \frac{A_{net}\,f_y}{\gamma_{M0}}.$$

3.2.2 Tension connections

In a tension connection the bolt is subjected to tension forces. Here, two sub-categories are defined:

a)   Category D: Non-preloaded

Preloading is not required and all bolt classes given in Table 3.1 may be used. This category may be used for connections subjected to predominantly static loading, for example wind loads acting on buildings. The design criteria to be checked are:

71

$$F_{t,Ed} \leq F_{t,Rd}$$
$$F_{t,Ed} \leq B_{p,Rd}$$

(3.6)

where:

$F_{t,Rd}$ is the design tension resistance of the bolt

$B_{p,Rd}$ is the design punching resistance of the bolt head and the nut.

b)   Category E: Preloaded

In preloaded tension connections, only bolt classes 8.8 and 10.9 should be used. This category relates to connections which are frequently subjected to variations of loading, for example in crane supporting structures. The design criteria are the same as for category D connection (non-preloaded), see Eq. (3.6). Note that, for preloaded bolts, controlled tightening according to EN 1090-2 is required.

When bolts are subjected to both shear and tension forces, interaction should also be checked, see 3.4.3.

If preloaded bolts are used in category A or category D connections where preloading is used to improve for example serviceability performance or durability or where preloading is required for execution purposes, but where preloading is not explicitly used in the design checks for slip resistance, then the level of preload can differ from the requirements in EN 1090-2. In this case, the National Annex can specify less strict requirements.

72

UK Specific Comment

The NA to BS EN 1993-1-8 states that if the preload is not explicitly used in the design calculations, then no specific level of preload is required.

3.3   POSITIONING OF BOLT HOLES

In bolted connections, independently of the type of loading, the bolt holes must fulfil certain requirements with respect to minimum and maximum spacing between the holes, end distances and edge distances. These requirements are given in Table 3.3 extracted from Table 3.3 of EN 1993-1-8. The symbols for the spacing and distances are defined in Figure 3.3. Note that these limits are valid for predominantly static loaded joints. For structures subjected to fatigue, requirements are given in EN 1993-1-9 (CEN, 2005d).

Table 3.3 – Minimum and maximum spacing, end and edge distances

Distances and spacings	Minimum	Maximum		
		Structures made from steels conforming to EN 10025 except steels conforming to EN 10025-5		Structures made from steels conforming to EN 10025-5
		Steel exposed to the weather or other corrosive influences	Steel not exposed to the weather or other corrosive influences	Steel used unprotected
End distance e_1	$1.2d_0$	$4t + 40$ mm		min $(8t, 125$ mm$)$
Edge distance e_2	$1.2d_0$	$4t + 40$ mm		min $(8t, 125$ mm$)$
Spacing p_1	$2.2d_0$	min $(14t, 200$ mm$)$		min $(14t_{min}, 175$ mm$)$
Spacing $p_{1,0}$		min $(14t, 200$ mm$)$		
Spacing $p_{1,i}$		min $(28t, 400$ mm$)$		
Spacing p_2	$2.4d_0$	min $(14t, 200$ mm$)$		min $(14t_{min}, 175$ mm$)$
where t is the thickness of the thinner outer connected part				

Maximum values are given to prevent corrosion in exposed members and to avoid local buckling between two fasteners in compression members. In other cases, or when no values are given, maximum values for the spacings, edge distances and end distances are unlimited. The local buckling resistance of the plate in compression between the fasteners should be calculated according to EN 1993-1-1 where a value of $0.6p_1$ should be used as buckling length. Local buckling between the fasteners need not to be checked if p_1/t is smaller than 9ε, with $\varepsilon = \sqrt{235/f_y}$. The edge distance should not exceed the local buckling requirements for an outstand element in the compression members, see EN 1993-1-1. The end distance is not affected by this requirement.

For slotted holes, minimum end distances and minimum edge distances are a little larger. For more information, see Table 3.3 of EN 1993-1-8.

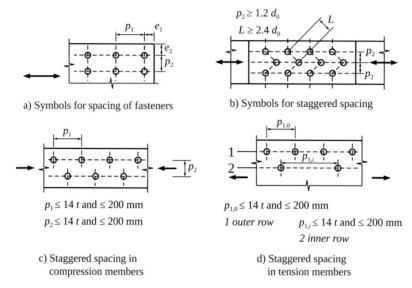

a) Symbols for spacing of fasteners

b) Symbols for staggered spacing

$p_1 \leq 14\,t$ and ≤ 200 mm
$p_2 \leq 14\,t$ and ≤ 200 mm

$p_{1,0} \leq 14\,t$ and ≤ 200 mm
1 outer row $p_{1,i} \leq 14\,t$ and ≤ 200 mm
             2 inner row

c) Staggered spacing in
compression members

d) Staggered spacing
in tension members

Figure 3.3 – Symbols for end and edge distances and spacing of fasteners

3.4 DESIGN OF THE BASIC COMPONENTS

3.4.1 Bolts in shear

If the shear plane passes through the *threaded portion* of the bolt, the design resistance of bolts in shear $F_{v,Rd}$ is

$$F_{v,Rd} = \frac{\alpha_v f_{ub} A_s}{\gamma_{M2}} \tag{3.7}$$

where:

α_v = 0.6 for (more ductile) bolt classes 4.6, 5.6 and 8.8
α_v = 0.5 for (less ductile) bolt classes 4.8, 5.8 and 6.8 and 10.9.

If the shear plane passes through the *shank* (unthreaded portion) of the bolt, the design resistance of bolts in shear $F_{v,Rd}$ is

$$F_{v,Rd} = \frac{0.6 f_{ub} A}{\gamma_{M2}}$$ (3.8)

The design shear resistance according to Eqs. (3.7) and (3.8) should only be used where the bolts are used in holes with nominal clearances as specified in EN 1090-2 (CEN, 2011), i.e. 1 mm for M12 or M14, 2 mm for M16 to M24 and 3 mm for M27 and thicker bolts. Table 3.4 summarises the values for the design shear resistance for common bolt diameters.

Table 3.4 – Design shear resistances $F_{v,Rd}$ in kN (for γ_{M2} = 1.25)

Shear plane	Bolt grade	M12	M16	M20	M24	M27	M30	M36
in the shank	4.6	21.71	38.60	60.32	86.86	109.9	135.7	195.4
	5.6	27.14	48.25	75.40	108.6	137.4	169.6	244.3
	8.8	43.43	77.21	120.6	173.7	219.9	271.4	390.9
	10.9	54.29	96.51	150.8	217.1	274.8	339.3	488.6
in the thread	4.6	16.19	30.14	47.04	67.78	88.13	107.7	156.9
	5.6	20.23	37.68	58.80	84.72	110.2	134.6	196.1
	8.8	32.37	60.29	94.08	135.6	176.3	215.4	313.7
	10.9	33.72	62.80	98.00	141.2	183.6	224.4	326.8

The design shear resistance for *fit bolts*, where the nominal diameter of the hole is the same as the nominal diameter of the bolt, i.e. holes with no clearance, should be determined using Eq. (3.8). The thread of a fit bolt should not be included in the shear plane.

3.4.2 Bolts in tension

The design tension resistance of a (non-preloaded or preloaded) bolt is determined according to EN 1993-1-8 as follows:

$$F_{t,Rd} = \frac{0.9 f_{ub} A_s}{\gamma_{M2}} \tag{3.9}$$

Table 3.5 gives the design tension resistances for common bolt diameters.

Table 3.5 – Design tension resistances $F_{t,Rd}$ in kN (for $\gamma_{M2} = 1.25$)

Bolt grade	M12	M16	M20	M24	M27	M30	M36
4.6	24.28	45.22	70.56	101.7	132.2	161.6	235.3
5.6	30.35	56.52	88.20	127.1	165.2	202.0	294.1
8.8	48.56	90.43	141.1	203.3	264.4	323.1	470.6
10.9	60.70	113.0	176.4	254.2	330.5	403.9	588.2

For countersunk bolts, in Eq. (3.9), a factor of 0.63 should be used instead of 0.9.

3.4.3 Bolts in shear and tension

When bolts are subjected to shear and tension forces, the following design criteria given in Table 3.4 of EN 1993-1-8 should be satisfied:

$$\frac{F_{v,Ed}}{F_{v,Rd}} + \frac{F_{t,Ed}}{1.4 F_{t,Rd}} \leq 1.0 \tag{3.10}$$

Figure 3.4 shows the interaction diagram for combined shear and tension. If the shear load in a bolt does not exceed about 28% of its shear resistance, the design tension resistance of a bolt must not be reduced. In other words, the interaction check is not needed, if

$$F_{v,Ed} \leq (1 - 1/1.4) F_{v,Rd} \approx 0.286 F_{v,Rd} \tag{3.11}$$

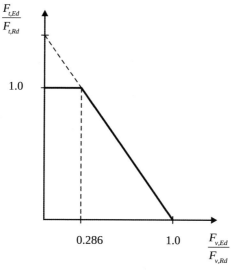

Figure 3.4 – Shear-tension interaction of bolts

3.4.4 Preloaded bolts

Depending on the performance specification for a connection, bolts may be used as preloaded or non-preloaded. In some particular cases, for example in connections subjected to dynamic loading, it is or it could be required that bolts are preloaded in order to ensure a minimum clamping force between connected plates. In such cases, a particular preparation of the surfaces as well as a controlled tightening is requested, see EN 1090-2. In these cases, the preload can be used in the design calculations, for example in slip resistant connections. The design preload $F_{p,Cd}$ is determined as follows:

$$F_{p,Cd} = \frac{0.7 f_{ub} A_s}{\gamma_{M7}}$$

(3.12)

However, procedures to control preloading are cost-intensive. Also, controlled preloading of bolts will decrease the speed of fabrication and erection of the steel structures. So, it is obvious that preloaded bolts with controlled tightening should only be used when it is absolutely necessary. On the other hand, if bolts are preloaded to some extent, even if the level of preloading is not controlled by a particular tightening procedure, this will improve the stiffness performance of a bolted connection, for example by decreasing deformations (for example for execution purposes). However, in this case, the preload cannot be used in the design calculations.

a)   Preloaded bolts in shear connections

The structural response in a connection is different depending on the load case. In a shear connection, as long as the bolts are not preloaded, the internal forces will be transferred by bearing between the plates and the bolt, and by shear in the bolt shank respectively, see Figure 3.5. If the bolts are preloaded, a clamping pressure will develop between the connected parts. The internal forces are transferred directly between the connected parts. Friction will prevent slipping of the connection, see Figure 3.5. This is called a slip resistant connection. In this case the resulting deformation of the connection is significantly smaller than the deformation of a bearing type connection.

Figure 3.5 – Load transfer in a non-preloaded and a preloaded connection in a shear connection

The maximum load which can be transferred by friction is the design slip resistance at the ULS $F_{s,Rd}$, which is dependent on the preload of the bolts $F_{p,C}$, and on the friction coefficient or slip factor of the clamped plates. The design slip resistance of a preloaded class 8.8 or 10.9 bolt should be taken as:

$$F_{s,Rd} = \frac{k_s \, n \, \mu}{\gamma_{M3}} F_{p,C} \qquad (3.13)$$

where:

k_s is a reduction factor shape and size of the hole; for normal holes $k_s = 1$, for other cases, e.g. oversized or slotted holes, reference is made to EN 1993-1-8

n is the number of the friction surfaces

μ is   the slip factor; values for the slip factor varies, depending on the treatment of friction surfaces, between 0.2 and 0.5. Slip factors μ and friction surface classes are defined in section 8.4 of EN 1090-2, see Table 3.6.

Values for the nominal preload force of the bolts $F_{p,C} = 0.7 f_{ub} A_s$ are given in Table 3.7.

Table 3.6 – Friction surface classes and slip factors

Surface treatment	Class	Slip factor
Surfaces blasted with shot or grit with loose rust removed, not pitted	A	0.5
Surfaces blasted with shot or grit: a) spray-metallized with an aluminium or zinc based product; b) with alkali-zinc silicate paint with a thickness of 50 to 80 µm	B	0.4
Surfaces cleaned by wire-brushing or flame cleaning, with loose rust removed	C	0.3
Surfaces as rolled	D	0.2

Table 3.7 – Nominal values of $F_{p,C}$ in kN

Bolt class	M12	M16	M20	M22	M24	M27	M30	M36
8.8	47	88	137	170	198	257	314	458
10.9	59	110	172	212	247	321	393	572

If the applied load exceeds the design slip resistance $F_{s,Rd}$, the connection transforms into a bearing type connection.

In order to make erection easier, bolt holes have a clearance with respect to the bolt diameter. In bearing type connections, hole clearances cause slips which lead to an extra deformation as illustrated in Figure 3.6 (Stark, 1995), whereas, in a connection with preloaded bolts, slip will only occur when the external load exceeds the slip resistance.

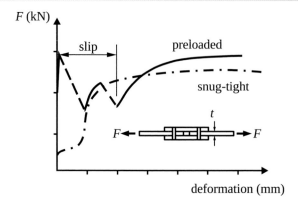

Figure 3.6 – Load-deformation diagram of a shear connection

Note that the slip resistance will decrease if the bolt is also subjected to an external tension load because the external tension load will reduce the clamping force. If a slip-resistant connection is subjected to an applied tensile force, $F_{t,Ed}$ or $F_{t,Ed,ser}$, in addition to the shear force, $F_{v,Ed}$ or $F_{v,Ed,ser}$, tending to produce slip, the design slip resistance per bolt should be taken as follows:

– for category B connections:

$$F_{s,Rd,ser} = \frac{k_s \, n \, \mu \left(F_{p,C} - 0.8 F_{t,Ed,ser} \right)}{\gamma_{M3}} \qquad (3.14)$$

– for category C connections:

$$F_{s,Rd} = \frac{k_s \, n \, \mu \left(F_{p,C} - 0.8 F_{t,Ed} \right)}{\gamma_{M3}} \qquad (3.15)$$

b)   Preloaded bolts in tension connections

In a connection where two or more plates are connected with preloaded bolts, the behaviour of the connection subjected to an external tension force is usually illustrated through the so-called force triangle, see Figure 3.7, where both the load-deformation diagrams of the bolt and of the plates are shown.

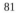

① initial stage
② preloaded stage
③ separation of plates
④ $F_e = F_t$

Figure 3.7 – Force triangle in a tension connection

Between stages ① and ②, the bolt is prestressed and the full preloading is reached at ②. The bolt and the plate package undergo the same deformation (bolt extends and plate package shortens). In addition, the tension force in the bolt and the clamping force between the plates are equal. An external load F_e is then applied. It causes an increase of the bolt elongation, but at the same time the plate package expands, what reduces the clamping force F_c. Hence, the external load is the sum of ΔF_t and ΔF_c (see Figure 3.7). Given that the stiffness of the plate package is usually much higher than the stiffness of the bolt, the additional force in the bolt is very small compared to the external load. The connected plates will separate at stage ③ for an external load which exceeds the preload of the bolt. This has a very positive effect on the fatigue behaviour of the connection and that is why it is recommended to avoid a separation of the plates in the case of fatigue loading.

c)   Tightening of preloaded bolts

Preloading of high strength structural bolts requires a controlled tightening. Appropriate methods are given in EN 1090-2:

- Torque method: the bolts shall be tightened using a torque wrench offering a suitable operating range. Hand or power operated wrenches may be used. The torque applied to the nut (or the head when the access to the nut side is inadequate) is used partly to overcome friction between the nut and the surface against which it rotates, and partly – approximately half – to drive the threads up the helix, overcoming the friction between the mating screw surfaces and the resolved component of the axial force. The tightening torque shall be applied continuously and smoothly. Tightening by the torque method comprises at least the two following steps:

- a first tightening step: the wrench shall be set to a torque value of about 75% of the torque reference values. This first step shall be completed for all bolts in one connection prior to commencement of the second step;
- a second tightening step: the wrench shall be set to a torque value of 110% of the torque reference values;

where the torque reference values must be declared by the fastener manufacturer in accordance with the relevant parts of EN 14399 (CEN, 2015) or determined by tests under site conditions according to the procedures given in Annex H of EN 1090-2.

- Combined method: tightening by the combined method comprises two steps:

- a first tightening step, using a torque wrench offering a suitable operating range (see also *torque method* described above). The wrench shall be set to a torque value of about 75% of the torque reference values. This first step shall be completed for all bolts in one connection prior to commencement of the second step;
- a second tightening step in which a specified partial turn is applied to the turned part of the assembly. The position of the nut relative to the bolt threads shall be marked after the first step, using a marking crayon or marking paint, so that the final rotation of the nut relative to the thread in this second step can be easily determined. Further rotation to be applied, during the second step of tightening depends on total thickness of parts to be connected. Values are given in Table 3.8 extracted from EN 1090-2. Where the surface under the bolt head or nut (allowing for example for tapered washers) is not perpendicular to the bolt axis, the required angle of rotation should be determined by testing.

Table 3.8 – Additional rotation for the combined method

Total nominal thickness t of parts to be connected (including all packs and washers)	Further rotation to be applied, during the second step of tightening	
	Degrees	Part turn
$t < 2d$	60	1/6
$2d \le t < 6d$	90	1/4
$6d \le t \le 10d$	120	1/3

Many other systems to control the preload in bolts have been tested. Most popular ones are "HRC systems" (bolt and nut assemblies with calibrated preload) and "direct tension indicators". Their use is very simple, but the scatter of the preload magnitude is large. They represent an alternative to the direct methods described before.

d)   HRC systems

HRC assemblies have a torque control groove and a torque control spline at the end of the threaded part as shown in Figure 3.8.

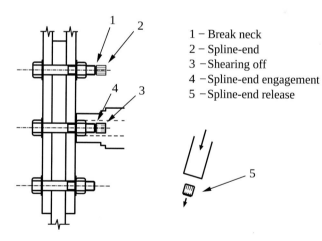

1 – Break neck
2 – Spline-end
3 – Shearing off
4 – Spline-end engagement
5 – Spline-end release

Figure 3.8 – HRC systems: Principle of tightening

When the torque reaches a definite value determined by the groove, the end of the screw breaks and the tightening is stopped. This type of bolt has advantages:

– it eliminates the need for a back-up man on the bolt screw.

– it removes the possibility of operator error and checking is very rapid.

– the tool requires no calibration and tightening is not affected by field conditions, at least as long as the requirements in terms of slip factor are satisfied.

On the other hand, the reduction in preload could be rather large. To avoid such an unexpected event, a new European standard EN 14399-10 has been introduced where requirements for HRC systems are stated (CEN, 2009b).

e)   Direct tension indicators

A direct tension indicator is an especially hardened washer with protrusions on one face, see Figure 3.9.

Figure 3.9 – Direct tension indicator

The protrusions bear against the underside of the bolt head leaving a gap. As the bolt is tightened the protrusions are flattened and the gap reduced. At a specified average gap, measured by feeler gauge, the induced shank tension will not be less than the minimum required by the standards. Figure 3.10 shows a standard assembly in place before and after tightening the bolt.

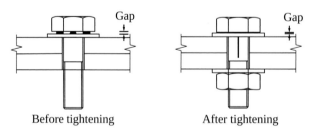

Before tightening   After tightening

Figure 3.10 – Principle of tightening with a direct tension indicator

All the bolts have to be tightened in accordance with the manufacturer's instructions and to the required tension indicator gap recommended by the manufacturer, as verified by calibration tests. Tightening shall be carried out progressively from the middle of the joint to the free edges, to avoid loosening previously tightened fasteners.

Tightening to the required gap shall be carried out in two stages to ensure consistency of preloading. In the first stage 75% of the deformation of the protrusions of the tension indicator shall be reached. In the second stage of final tightening the required gap shall be reached.

Bolt and nut assemblies with direct tension indicators are standardised in EN 14399-9 (CEN, 2009a).

3.4.5 Plates in bearing

In bearing type connections, the internal force in the plate is transferred to the bolt by hole bearing. The design bearing resistance is determined as follows:

$$F_{b,Rd} = \frac{k_1 \alpha_b f_u d t}{\gamma_{M2}} \tag{3.16}$$

where:

k_1 = $\min\left(2.8 e_2/d_0 - 1.7; 1.4 p_2/d_0 - 1.7; 2.5\right)$ for edge bolts and $\min\left(1.4 p_2/d_0 - 1.7; 2.5\right)$ for inner bolts

α_b = $\min\left(\alpha_d; f_{ub}/f_u; 1.0\right)$

         where $\alpha_d = e_1 / 3d_0$ for end bolts and $\alpha_d = p_1 / 3d_0 - 0.25$ for inner bolts

It should be noted that Eq. (3.16) covers two failure modes, see Figure 3.11. For small end distances or small bolt pitches in the direction of load transfer, the plate can fail by plate shear (Figure 3.11a). Otherwise it will fail by elongation of the hole (Figure 3.11b). Plate shear failure is taken into account by the factor α_d (Figure 3.11c). The factor k_1 considers that the bearing resistance is reduced if the edge distance e_2 or hole distance p_2 perpendicular to the direction of load transfer is small. If e_2 and p_2 are large enough, the full bearing capacity can develop.

| Plate shear | Hole elongation | Simplified model for shear |
| Mode of failure | of the plate | failure of the plate |

Figure 3.11 – Failure modes for a plate in bearing

The bearing resistance is calculated using the ultimate resistance of the plate. However, when the ultimate strength of the bolt is weaker than the ultimate strength of the plate, the bearing resistance is reduced to account for this situation.

Note that in EN 1993-1-8 this component is called "bolts in bearing on beam flange, column flange, end plate and cleat".

3.4.6 Block tearing

Block tearing failure, also called block shear failure can occur at a group of fastener holes, for example near the end of a beam web or a bracket. Figure 3.12 shows different situations of block tearing. It may be prevented by using appropriate hole spacing. Block tearing generally consists of tensile rupture along the line of fasteners on the tension face of the hole group and by yielding in shear at the row of fastener holes along the shear face of the hole group, see Figure 3.12.

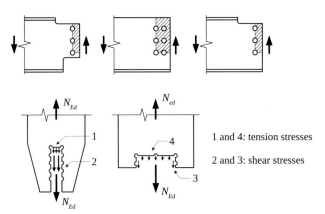

1 and 4: tension stresses

2 and 3: shear stresses

Figure 3.12 – Block tearing failure

The design block tearing resistance is for a symmetric bolt group

– subject to concentric loading, $V_{eff,1,Rd}$:

$$V_{eff,1,Rd} = \frac{A_{nt}f_u}{\gamma_{M2}} + \frac{A_{nv}f_y}{\sqrt{3}\gamma_{M0}}$$

(3.17)

– subject to eccentric loading, $V_{eff,2,Rd}$:

$$V_{eff,2,Rd} = \frac{0.5A_{nt}f_u}{\gamma_{M2}} + \frac{A_{nv}f_y}{\sqrt{3}\gamma_{M0}}$$

(3.18)

where:

A_{nt} is net area subjected to tension

A_{nv} is net area subjected to shear.

3.4.7 Injection bolts

Injection bolts are bolts in which the cavity resulting from the clearance between the bolt shank and the wall of the plate is completely filled up with a two component resin (ECCS, 1994). After tightening the bolt, the resin in "injected" in the cavity through a little hole in the head of the bolt, see Figure 3.13 (ECCS, 1994). The resin, after complete curing, has a defined minimum resistance and the connection becomes a slip resistant connection.

Figure 3.13 – Injection bolts in a double lap joint

EN 1993-1-8 provides design rules for injection bolts. These should be of class 8.8 or 10.9 and may be non-preloaded or preloaded, i.e. bolts can be designed for Category A, B or C connections. For a connection with injection bolts, the design bearing resistance is taken as the design bearing resistance of the resin:

$$F_{b,Rd,resin} = \frac{k_t \, k_s \, d \, t_{b,resin} \beta f_{b,resin}}{\gamma_{M4}}$$ (3.19)

where:

k_t is 1.0 for SLS (long duration) or 1.2 for ULS

k_s is taken as 1.0 for holes with normal clearances, for oversized holes see 3.6.2.2 of EN 1993-1-8

$t_{b,resin}$ is the effective bearing thickness of the resin, see Table 3.9

β is a coefficient depending of the thickness ratio of the connected plates, see Table 3.9

$f_{b,resin}$ is the bearing strength of the resin. No values are given in EN 1993-1-8. Reference is made to EN 1090-2. Annex K of EN 1090-2 states that the design bearing strength of the resin should be determined similar to the procedure for the determination of the slip factor as specified in Annex G of EN 1090-2.

Table 3.9 – Values for $t_{b,resin}$ and β

t_1/t_2	$t_{b,resin}$	β
≥ 2.0	$2t_2 \leq 1.5d$	1.0
$1.0 < t_1/t_2 < 2.0$	$t_1 \leq 1.5d$	$1.66 - 0.33\, t_1/t_2$
≤ 1.0	$t_1 \leq 1.5d$	1.33

Experimental tests (Gresnigt *et al*, 2000) have demonstrated that connections made with injection bolts have a very good long time behaviour. After a quite moderate relaxation of the resin, the deformation of bearing connections remains more or less constant. Even if the fabrication and erection of such connections requires a special attention, see Annex K of EN 1090-2, this type of fasteners could be an alternative of fitted bolts or rivets in order to fabricate a slip resistant bolted connection without using preloaded bolts.

UK Specific Comment

Injection bolts are rarely used in the UK.

3.4.8 Pins

Pins are normally used in lap joints only subjected to shear forces, such as tension rods or pin-ended columns. A solid circular pin is placed through the holes of the connected elements. In contrast to a normal bolt, a pin has no head and no threaded part of its shank. It is only secured to ensure that the pin will not become loose. As the pin has no head or nut, the connected plates will not necessary be clamped together and a gap may remain between the connected plates. Therefore, pin-ended members can behave as a perfect hinge allowing for significant rotations. On the other hand, bending moments that may rise in the pin need to be taken into account in the design.

In pin-connected members the geometry of the unstiffened element that contains a hole for the pin should satisfy the dimensional requirements given in Figure 3.14.

Type A: Given thickness t

$$a \geq \frac{F_{Ed}\gamma_{M0}}{2t\,f_y} + \frac{2d_0}{3}$$

and

$$c \geq \frac{F_{Ed}\gamma_{M0}}{2t\,f_y} + \frac{d_0}{3}$$

Type B: Given geometry

$$t \geq 0.7\sqrt{\frac{F_{Ed}\gamma_{M0}}{f_y}}$$

and

$$d_0 \leq 2.5t$$

Figure 3.14 – Geometrical requirements for pin ended members

Furthermore, pin-connected members should be arranged to avoid any eccentricity and should be of sufficient size to distribute the load that part of the member in the region of the pin hole into the main body of the member.

Design rules can be found in Section 3.13 of EN 1993-1-8. They are reproduced here below. As a single exception, EN 1993-1-8 also allows the design of a pin as a single bolted connection as long as no significant rotation is required and the length of the pin is less than three times its diameter. Trusses constitute a potential application field for such a "bolt behaviour".

The design requirements for solid circular pins are given in Table 3.10. The moments in the pin are calculated on the basis that the connected parts form simple supports and it is generally assumed that the reactions between the pin and the connected parts are uniformly distributed along the length in contact on each part as indicated in Figure 3.15.

Table 3.10 – Design formulae for pin connections

Failure mode	Design requirements
Shear resistance of the pin	$F_{v,Rd} = 0.6\, A f_{up} / \gamma_{M2} \geq F_{v,Ed}$
Bearing resistance of the plate and the pin	$F_{b,Rd} = 1.5\, t\, d f_y / \gamma_{M0} \geq F_{b,Ed}$
If the pin is intended to be replaceable this requirement should also be satisfied.	$F_{b,Rd,ser} = 0.6\, t\, d f_y / \gamma_{M6,ser} \geq F_{b,Ed,ser}$
Bending resistance of the pin	$M_{Rd} = 1.5\, W_{el} f_{yp} / \gamma_{M0} \geq M_{Ed}$
If the pin is intended to be replaceable this requirement should also be satisfied.	$M_{Rd,ser} = 0.8\, W_{el} f_{yp} / \gamma_{M6,ser} \geq M_{Ed,ser}$
Combined shear and bending resistance of the pin	$\left[\dfrac{M_{Ed}}{M_{Rd}}\right]^2 + \left[\dfrac{F_{v,Ed}}{F_{v,Rd}}\right]^2 \leq 1$

d	is	the diameter of the pin;
f_y	is	the lower of the design strengths of the pin and the connected part;
f_{up}	is	the ultimate tensile strength of the pin;
f_{yp}	is	the yield strength of the pin;
t	is	the thickness of the connected part;
A	is	cross-sectional area of the pin.

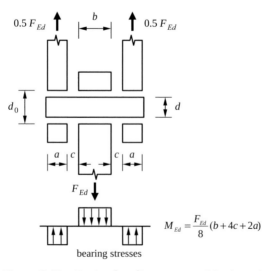

Figure 3.15 – Design bending moment M_{Ed} in a pin

If the pin is intended to be replaceable, in addition to the provisions given before, the contact bearing stress should satisfy:

$$\sigma_{h,Ed} \le f_{h,Rd} \tag{3.20}$$

with:

$$\sigma_{h,Ed} = 0.591 \sqrt{\frac{EF_{b,Ed,ser}(d_0 - d)}{d^2 t}} \tag{3.21}$$

$$f_{h,Rd} = 2.5 \frac{f_y}{\gamma_{M6,ser}} \tag{3.22}$$

In these equations:

d is   the diameter of the pin;

d_0 is   the diameter of the pin hole;

$F_{b,Ed,ser}$ is   the design value of the force to be transferred in bearing, under the characteristic load combination for SLS.

UK Specific Comment

Reference is made to SCI (1995b) to calculate the bending moments in the pin of a pin bearing.

3.4.9 Blind bolting

3.4.9.1 Flow drill blind bolting

The flow drill system is a special patented method for extruded holes. Flow drilling is a thermal drilling process to make a hole through a steel plate, for example the wall of hollow section as shown in Figure 3.16, by bringing a tungsten carbide bit into contact with steel plate and generating sufficient heat by friction to soften the steel. The bit moves through the wall, the metal flows to form an internal bush. In the subsequent step, the bush is threaded using a roll tap (Kurobane *et al*, 2004). Flow drill process is illustrated in Figure 3.17.

Figure 3.16 – Flow drill connection for joining end plates to RHS

Note that flow drill blind bolting is not mentioned in EN 1993-1-8. It must be checked with the local building authorities if this type of bolting can be used when a structure is designed to EN 1993.

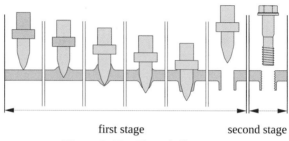

first stage   second stage

Figure 3.17 – Flow drill process

Kurobane *et al* (2004) provides practical guidelines on flow drill blind bolting:

- wall thicknesses up to 12.5 mm can be recommended by using the flow drill method;
- threaded roll tapped holes for bolt sizes M16, M20 and M24 can be made;
- the full tension capacity of grade 8.8 bolts can be carried, provided that the plate thickness is equal to or greater than the minimum thickness shown in Table 3.11 and f_y is between 275 N/mm^2 and 355 N/mm^2;

Table 3.11 – Minimum plate thickness

Bolt size	Minimum plate thickness [mm]
M16 grade 8.8	6.4
M20 grade 8.8	8.0
M24 grade 8.8	9.6

   – the shear and bearing capacities of the hole can be calculated in the normal manner;

   – In most applications in which the bolts are loaded in tension, the deformation or yielding of the steel plate will determine the connection capacity and not the capacity of each individual bolt.

For more details the reader may consult Kurobane *et al* (2004).

3.4.9.2 SHS Blind bolting connections

In recent years, other systems than the flow drill ones have been developed for connections with one side access only. A number of patented blind bolting systems is available, e.g. Huck "Ultra Twist Blind Bolt", Lindapter "HolloFast", see Figure 3.18 (Wardenier *et al*, 2010) and "HolloBolt".

Figure 3.18 – Lindapter "HolloFast"

The systems are based on the principle that after bringing them in from one side the bolts are torqued and a "bolt head" forms on the inside of the connection.

The design rules for blind bolting systems are based on typical failure modes, i.e.:

   – Punching shear of the fastener through the column face;
   – Yielding of the column face;
   – Bolt failure in shear, tension or a combination of both.

By means of these different systems, beams with end plates are directly connected to column faces by tension bolts. Sometimes, the column walls have to be reinforced to prevent local distortion of the column walls or when full strength connections are required. The column walls may be partially thickened over the areas where the end plates are attached.

UK Specific Comment

Flow drill blind bolting is not common practice in the UK. SHS blind bolting is a common system, but there are several manufactures other than that suggested in the book.

3.4.10 Nails

In some cases nails can be used to form reliable structural connections; for example, Circular Hollow Sections (CHS) can be nailed together as illustrated in Figure 3.19 (Wardenier *et al*, 2010) for a splice connection between two co-axial tubes. In such a connection, one tube can fit snugly inside the other.

Figure 3.19 – Nailed connection

The observed failure modes are:

– Nail shear failure;
– Tube bearing failure;
– Net section fracture of the tube.

Simple design formulae originally derived for bolted and riveted connections, are proposed in Packer (1996).

> **UK Specific Comment**
>
> Nails are rarely used in the UK.

3.4.11 Eccentricity of angles

Joints have to transfer the internal forces, obtained by means of a structural frame analysis, between the frame members. Local eccentricities in the joints may however generate additional internal forces. The joints and members should be designed for the total resulting moments and forces, except in the case of particular types of structures where it has been demonstrated that it is not necessary; for instance in lattice girders, provided some conditions are met (see EN 1993-1-8).

One of the common situations where such eccentricities are to be considered is the case of an angle connected on one leg only by one or more lines of bolts; in-plane and out-of-plane eccentricities are here identified by considering the relative positions of the centroidal axis of the member and of the setting out line (see Figure 3.20).

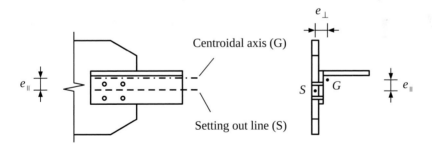

$e_\parallel$: eccentricity parallel to the gusset

$e_\perp$: eccentricity perpendicular to the gusset

Figure 3.20 – Local eccentricities in a bolted angle

For the specific case of angles connected through one line of bolts and subjected to an axial tension force only, EN 1993-1-8 provides a simplified design method which consists in virtually replacing the actual member by a concentrically loaded one (i.e. loaded along the centroidal axis), but characterised by an effective area. The design resistance of the member in the net section is then defined as follows (Figure 3.21):

95

one bolt:
$$N_{u,Rd} = \frac{2.0\left(e_2 - 0.5d_0\right)tf_u}{\gamma_{M2}}$$
(3.23)

two bolts in line:
$$N_{u,Rd} = \frac{\beta_2 A_{net} f_u}{\gamma_{M2}}$$
(3.24)

three bolts in line:
$$N_{u,Rd} = \frac{\beta_3 A_{net} f_u}{\gamma_{M2}}$$
(3.25)

In these equations:

β_2, β_3 are reduction factors dependent on the pitch p_1 distance as given in Table 3.12. For intermediate values of p_1, the value of β may be determined by linear interpolation;

A_{net} is the net area of the angle. For an unequal-leg angle connected by its smaller leg, A_{net} should be taken as equal to the net section area of an equivalent equal-leg angle of leg size equal to that of the smaller leg.

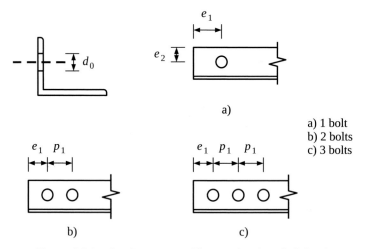

a)

a) 1 bolt
b) 2 bolts
c) 3 bolts

b)

c)

Figure 3.21 – Angle connected by one leg (one bolt line)

Table 3.12 – Reduction factors β_2 and β_3

Pitch	p_1	$\leq 2.5d_0$	$\geq 5.0d_0$
2 bolts	β_2	0.4	0.7
3 bolts or more	β_3	0.5	0.7

3.5   DESIGN OF CONNECTIONS

3.5.1 Bolted lap joints

3.5.1.1 Introduction

The term "lap joint" designates a basic connection aimed at transferring forces between plates subjected to tension. One usually distinguishes between single and double overlap joints (see Figure 3.22). To achieve the force transfer, various connection elements may be used; but in practice, bolts and welds are the most commonly used, see Figure 3.23.

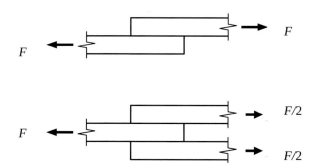

Figure 3.22 – Single and double overlap joints (lap joints)

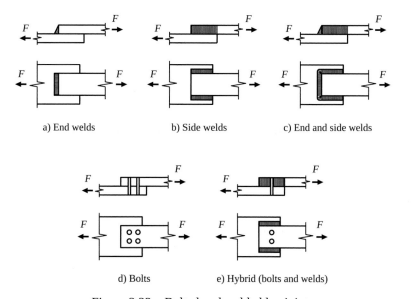

a) End welds   b) Side welds   c) End and side welds

d) Bolts   e) Hybrid (bolts and welds)

Figure 3.23 – Bolted and welded lap joints

In the next sections, the behaviour and the design of bolted lap joints are discussed. For welded and hybrid lap joints, reference should be respectively made to sub-chapter 4.4 and section 1.6.2.

3.5.1.2 Joints with non-preloaded bolts

In such joints, the forces are transferred from one plate to the other(s) by bolt-plate contact (Figure 3.24).

Figure 3.24 – Single and double bolted overlap joints

If the small friction effects between the plates and the bending forces in the bolts are neglected, the resistance of the joint may be limited by one of the following components:

– The bearing resistance of the plates and/or bolts;
– The shear resistance of the bolts (one or two shear planes, respectively for single or double overlap joints);
– The tension resistance of the plates.

The design resistance of a bolt in shear and of a plate/bolt in bearing has been defined in sections 3.4.1 and 3.4.5, respectively. For the design resistance of a plate in tension, reference is made to EN 1993-1-1 or to the ECCS Design Manual on "Design of Steel Structures" (Simões da Silva *et al*, 2010) where expressions are provided for the design of the gross and net sections.

Locally, i.e. around a bolt, the force that may be transferred can never exceed the design resistance of the weaker of the two first here-above mentioned components.

As far as the whole joint is concerned (with more than one bolt row), the force transferred between the plates is known to be non-uniformly distributed amongst the bolt rows (see distribution of bolt forces C_i in Figure 3.25a) (Ju *et al*, 2004). If sufficient deformation is provided around each bolt zone, a full plastic redistribution of forces may occur, otherwise failure is reached by lack of ductility and the maximum external force to be transferred is lower than the one corresponding to a full plastic distribution. Schematically, the different stages of forces distribution in a shear bolted connection may be represented as in Figure 3.25.

a) None of the bolt rows yields

b) Outer bolt rows reach yielding (elastic resitance of the connection)

99

c) Progressive yielding of bolt rows from outside to inside (plastic redistribution)

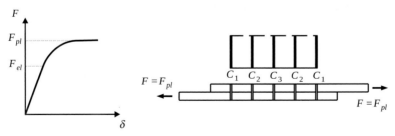

d) All bolt rows yield (full plastic resistance of the connection)

Figure 3.25 – Different stages of bolt force distribution in shear bolted connections

In Eurocode 3 Part 1-8, the design resistance of the lap joint is defined as the smaller of the two following values:

- – The design resistance of the weakest plate(s) in tension;
- – The design resistance of the group of fasteners (the bolts).

The design resistance of the group of fasteners may be taken as the sum of the design bearing resistances $F_{b,Rd}$ of the individual fasteners provided that the design shear resistance $F_{v,Rd}$ of each individual fastener (rather brittle failure mode) is greater than or equal to the local design bearing resistance $F_{b,Rd}$ (rather ductile failure mode). Fulfilment of this criterion guarantees a sufficient deformation capacity to achieve a full redistribution of forces between the fasteners. The plastic resistance of the joint may so be reached.

Otherwise the design resistance of a group of fasteners should be taken as the number of fasteners multiplied by the smallest design resistance of any of the individual fasteners, what may be considered as a way to limit the plastic redistribution of yielding along the bolt rows.

Besides that, EN 1993 does not permit a full redistribution of plasticity in the case of so-called "long joints". This aspect is addressed in section 3.5.4.

3.5.1.3 Joints with preloaded bolts

Two categories of connections are considered:

- – Slip resistance at ULS (Category C);
- – Slip resistance at SLS (Category B).

In the case of Category B connections, the design resistance of the lap joint is evaluated as specified in section 3.5.1.1.

For Category C, the design resistance is equal to the total slip resistance, which is obtained by multiplying the design slip resistance of a bolt by the number of bolts in the joints, see Eq. (3.13).

As already mentioned in section 3.2.1c), the resistance of the net cross section of the plates at the bolt holes has to be checked in a different way for category C joints, see Eq. (3.5).

3.5.2 Bolted T-stubs

3.5.2.1 Generalities

The transfer of tensile forces in bolted joints is often achieved through the following basic components: column flange in bending, end plate in bending or cleat leg in bending, i.e. plates subjected to transverse forces.

The evaluation of the resistance of such plate components is based on a geometrical idealisation of the tension zone, known as the "T-stub" idealisation. A T-stub is a T profile made of a web in tension, and a flange in bending bolted to a rigid support, see Figure 3.26.

The response of the actual plate components is evaluated through the study of corresponding equivalent T-stubs; the equivalence is reached through the definition of an appropriate length of the equivalent T-stubs, called l_{eff}.

The use of T-stubs has the advantage that the different plate components acting in bending may be studied using the same basic model and thus with the same design formulae, whatever the mechanical property being considered (stiffness, resistance, ductility).

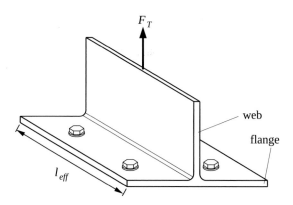

Figure 3.26 – T-stub geometry

Figure 3.27 – Visualisation of equivalent T-stubs in bolted connections

3.5.2.2 Design resistance

Zoetemeijer (1974) initially developed the T-stub model for unstiffened plated components and then extended it to some stiffened configurations. Later, Jaspart (1991) applied the concept to cleated connections before extending it again further to various plate configurations (Jaspart, 1997).

In terms of resistance, three different failure modes may be considered according to the geometry and mechanical response of the constitutive plates and bolts, see Figure 3.28:

– Failure mode 1: Yielding of the flange

   The resistance is associated to the formation of a plastic yield mechanism in the flange. In such a case, the bolts are sufficiently strong to resist to the applied axial tension forces, including the prying forces Q.

– Failure mode 2: Failure of the bolts further to a partial yielding of the flange

   Mixed failure is achieved through the formation of yield lines in the flange (the full plastic mechanism being not reached) and the failure of the bolts in tension (again including prying effects).

– Failure mode 3: Failure of the bolts

   The resistance is linked to the failure of the bolts in tension. The deformation of the flange in bending is small, resulting in an absence of prying effects.

Basic component	Equivalent T-stub	Moment distribution

Failure mode 1: Yielding of the flange

$F_{T,Rd}$

Prying force Q Q

$Q+0.5F_{T,Rd} < 0.5\sum F_{t,Rd}$

$M_{pl,Rd}$

$M_{pl,Rd} = l_{eff}\, m_{pl,Rd}$ $M_{pl,Rd}$

Failure mode 2: Failure of the bolts further to a partial yielding of the flange

$F_{T,Rd}$

Q Q

$0.5\sum F_{t,Rd}$

$M_{pl,Rd}$

Failure mode 3: Failure of the bolts

$F_{T,Rd}$

$0.5\sum F_{t,Rd}$

M_{Ed}

$M_{ed} \leq M_{pl,Rd}$

Figure 3.28 – Failure modes in the actual component and in the equivalent T-stubs

Failure mode 3   Failure mode 1   Failure mode 2

Figure 3.29 – Failure modes of an equivalent T-stub

These failure modes occur similarly in both the actual component and in the equivalent T-stub. The effective length l_{eff} is defined such that the resistance of the T-stub and of the actual component are the same. The evaluation of the resistance of the actual component may therefore be replaced by the study of the equivalent T-stub. As will be shown later, the T-stub model can also be used as a reference in determining the stiffness of the basic components.

EN 1993-1-8 defines the design resistance of the T-stub, for each failure mode, as follows (Figure 3.29):

- Mode 1: Yielding of the flange

$$F_{T,Rd,1} = \frac{4\,l_{eff,1}\,m_{pl,Rd}}{m} \tag{3.26}$$

- Mode 3: Failure of the bolts

$$F_{T,Rd,3} = \sum F_{t,Rd} \tag{3.27}$$

- Mode 2: Failure of the bolts further to partial yielding of the flange

$$F_{T,Rd,2} = \frac{2\,l_{eff,2}\,m_{pl,Rd} + n\sum F_{t,Rd}}{m+n} \tag{3.28}$$

where:

$m_{pl,Rd} = \dfrac{t^2 f_y}{4\gamma_{M0}}$ is the plastic bending resistant of the T-stub flange per

unit length

$\sum F_{t,Rd}$ is the sum of the design resistances $F_{t,Rd} = \dfrac{0.9\,A_s\,f_{ub}}{\gamma_{M2}}$ of

all the bolts in the T-stub

m, e are geometrical dimensions characterising the position of the potential yield lines in the T-stub flange (see Figure 3.29)

n is   a geometrical dimension characterising the position of the prying force: $n = e_{min}$, but with a limitation to $1.25m$; e_{min} is the smaller edge distance if two connected plates have different widths as show in Figure 3.30

$l_{eff,1}$, $l_{eff,2}$ minimum effective lengths of the yield lines for all the possible mechanisms in the T-stub, according to the failure modes.

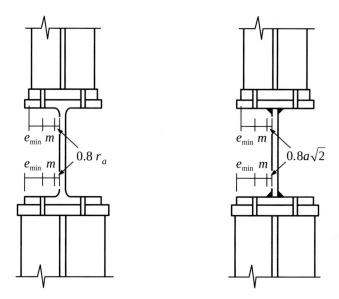

Figure 3.30 – Definition of e_{min} (for example in a beam-to-column joint)

In Figure 3.31a, the so-called *individual non-circular* yield line *patterns* are illustrated. In other cases, depending on the geometry of the component, *individual circular patterns* could also appear (Figure 3.31b). Different values of effective lengths are associated to these two types of yield patterns.

The formation of non-circular yield patterns requires the development of prying forces in the T-stub. Failure modes 1, 2 or 3 may therefore occur, depending on the value of β, as shown in Figure 3.32. On the contrary, when circular patterns form, no prying forces develop in the T-stub and failure mode 2 can not occur. In this case, a direct transition from failure mode 1 to failure mode 3 is assumed (Figure 3.32).

a) Individual non circular yield patterns for the resistance of T-stubs with a single bolt-row

b) Individual circular yield patterns for the resistance of T-stubs with a single bolt-row

c) Combined non circular yield patterns for the resistance of a group of bolt-rows

Figure 3.31 – Possible yield line mechanisms

$\lambda = n/m$

——— Non circular patterns

– – – Circular patterns

$$\beta = \frac{4m_{pl,Rd}\,(l_{eff,1}\ or\ l_{eff,2})}{m\Sigma F_{t,Rd}}$$

Figure 3.32 – Type of failure according to the geometry of the T-stub

The design resistance of the T-stub is the lowest of the two following values:

$$- \quad F_{T,Rd,A} = \min\left(F_{T,Rd,1}; F_{T,Rd,2}; F_{T,Rd,3}\right) \quad \text{for non-circular patterns} \qquad (3.29)$$

$$- \quad F_{T,Rd,B} = \min\left(F_{T,Rd,1}; F_{T,Rd,3}\right) \quad \text{for circular patterns} \qquad (3.30)$$

Therefore:

$$F_{T,Rd} = \min\left(F_{T,Rd,A}; F_{T,Rd,B}\right) \qquad (3.31)$$

In addition to considering T-stubs with a single bolt row, as individual, the possibility exists to form yield line mechanism involving adjacent bolt rows. Figure 3.31c illustrates the case of non-circular yield patterns extending over three bolt rows. The same may also apply to circular patterns. In practice, the resistance of the T-stub involving a group of bolt rows will always have to be compared to the sum of the design resistances of the individual T-stubs (involving one bolt row only) and the lower of these two resistances is used in the design.

Note that the resistance formulae in Eqs. (3.26) to (3.28) are only valid for a T-stub with two bolts per row. However, in practice, also configurations with four bolts per row are used. An extension of the design methods described in this section to a T-stub with four bolts per row can be found in Demonceau *et al* (2011).

107

3.5.2.3 *Influence of the actual bolt dimensions on the design resistance*

In Mode 1, a full plastic mechanism forms in the T-stub flange:

- Two interior longitudinal yield lines at the toe of the root radius (or radius of fillet in a welded T-stub);
- Two exterior longitudinal yield lines close to the bolts.

If the bolt force is assumed to be applied to the T-stub flange as a point load, as illustrated in Figure 3.33a, the exterior yield lines are straight and pass through the centre line of the bolts. Eq. (3.26) above covers this situation.

If the bolts were infinitely stiff (Figure 3.33c), the two exterior yield lines would "move" towards the T-stub web, resulting in a decrease of the

m distance and therefore in an increase of the design resistance $F_{T,Rd,1}$. However the reality is somewhat different and Figure 3.33b shows that the actual yield lines will be affected by the geometrical dimensions of the screw head and washer, by the rigidity of the bolt and by the degree of bolt preloading. For the practical application of the T-stub model, the determination of the actual yield lines and the evaluation of the corresponding T-stub resistance would be too complicated and time-consuming. EN 1993-1-8 therefore suggests an alternative method in which (Figure 3.33c):

- The assumption of straight yield lines passing through the bolt axes is kept;
- The positive local effect of the bolt head is considered by assuming that the bolt force is transferred to the T-stub flange in a distributed way, under the bolt head/washer (over a d_w width).

The distributed load on the T-stub, applied on a d_w width, may be further replaced by two concentrated forces acting at a e_w distance of the bolt line. The value of e_w is defined as follows:

$$e_w = \frac{d_w}{4} \qquad (3.32)$$

where:

d_w is the diameter of the washer, or the width across points of the head or nut, as relevant.

The refined alternative formula for Mode 1 writes then:

$$F_{T,Rd,1} = \frac{(8n - 2e_w)\, l_{eff,1}\, m_{pl,Rd}}{2\,m\,n - e_w(m+n)} \qquad (3.33)$$

and gives higher results than Eq. (3.26), and has no specific limit of application. Jaspart (1991) who developed the formulae also pointed out the possible benefit which would result, in addition, from consideration of the preloading of the bolt. This extra benefit is presently not considered by EN 1993-1-8.

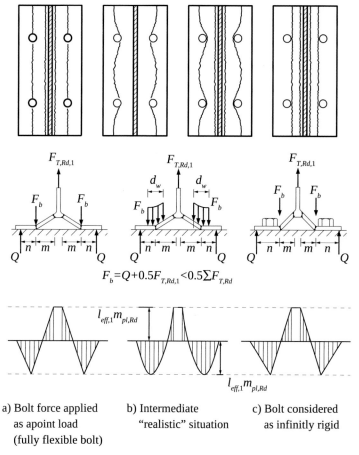

$$F_b = Q + 0.5F_{T,Rd,1} < 0.5\Sigma F_{T,Rd}$$

a) Bolt force applied as apoint load (fully flexible bolt)

b) Intermediate "realistic" situation

c) Bolt considered as infinitly rigid

Figure 3.33 – Influence of the bolt geometry on the yield lines

3.5.2.4 Influence of the bolt/anchor length

In the previous sections, except for Mode 3 failure, prying forces develop at the interface between the T-stub flange and its foundation. This leads to overloading of the bolts in tension.

However in the case where the bolts would elongate to a significant extent, the prying forces could no more physically develop. In particular this may occur in column bases where the elongation of the anchor bolts could more than compensate the flexural deformation of the plate.

As a direct consequence, the validity of the design formulae presented in sections 3.5.2.2 and 3.5.2.3 becomes questionable. In EN 1993-1-8, a criterion has been included so as to identify situations where prying forces develop:

$$L_b \leq \frac{8.8m^3 A_s n_b}{\sum l_{eff,1} t^3} \tag{3.34}$$

where:

L_b is the bolt elongation length, taken equal to the grip length (total thickness of material and washers), plus half the sum of the height of the bolt head and the height of the nut

or

the anchor bolt elongation length, taken equal to the sum of 8 times the nominal bolt diameter, the grout layer, the plate thickness, the washer and half the height of the nut

n_b is the number of bolt rows (with 2 bolts per row).

When this criterion is not satisfied, Mode 1 and Mode 2 design formulae have to be substituted by a single design formula:

$$F_{T,1-2,Rd} = \frac{2M_{pl,1,Rd}}{m} \tag{3.35}$$

Note: Eq. (3.35) in the present version of EN 1993-1-8 is safe but not correct. In Table 6.2 of EN 1993-1-8, the value $M_{pl,1,Rd}$ is defined as

$$M_{pl,1,Rd} = \frac{0.25 \Sigma l_{eff,1} t_f^2 f_y}{\gamma_{M0}} \tag{3.36}$$

where $l_{eff,1}$ is the effective length for Mode 1, i.e. the minimum of $l_{eff,cp}$ for circular patterns and $l_{eff,np}$ for non-circular patterns whereas, in reality, only the patterns developing without prying forces need to be checked.

3.5.2.5 Direct applications to flange plate connections

3.5.2.5.1 RHS flange-plate connections in tension

A direct application of the T-stub model may be considered for various Rectangular Hollow Sections (RHS) flange-plate connections in tension, see Figure 3.34a. The idealisation of the connection by means of an equivalent T-stub is illustrated in Figure 3.34b. The evaluation of the tension resistance of the connection therefore simply resumes to the determination of the T-stub resistance.

A limitation however exists which may be expressed as follows: the distance d (Figure 3.34) between the two extreme bolts in a row should not exceed the width of the RHS profile. If this condition is not fulfilled, specific yield mechanisms will appear in the plate corners, possibly involving the lateral sides of the plate. As a consequence, the T-stub, where only the bolted sides of the plate are considered, would not reflect the actual response of the connection.

Indication on how to overcome this "d distance" limitation may be found in Demonceau *et al* (2012).

a) Connection layout

b) Substitution by an equivalent T-stub

Figure 3.34 – RHS flange-plate connection in tension (plate bolted on two sides)

Another example is given in Figure 3.35a, where the plate is bolted on the four sides of the RHS profile. Similarly to what is said above, the connection may be modelled here by two equivalent T-stubs; and, in the case of a Square Hollow Section (SHS) profile, the resistance of the connection is basically the sum of the resistance of the two T-stubs.

Should however the actual yield line patterns developing in the four extended parts of the plate merge into a single common line (see an example in Figure 3.35b) appropriate values of the effective lengths should be considered when evaluating the T-stub resistances.

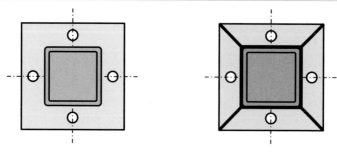

a) Connection layout   b) Possible combined yield line pattern

Figure 3.35 – SHS flange-plate connection in tension (plate bolted on four sides)

3.5.2.5.2 CHS flange-plate connections in tension

In fact, EN 1993-1-8 is not directly proposing recommendations for the design of bolted joints with circular flanges (Figure 3.36).

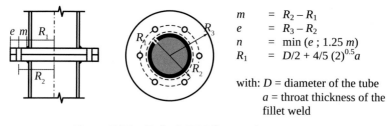

$$
\begin{aligned}
m &= R_2 - R_1 \\
e &= R_3 - R_2 \\
n &= \min (e\ ;\ 1.25\ m) \\
R_1 &= D/2 + 4/5\ (2)^{0.5}a
\end{aligned}
$$

with: D = diameter of the tube
a = throat thickness of the fillet weld

Figure 3.36 – Bolted CHS flange-plate connection

However, in Hoang *et al* (2013), the concept of T-stub has been extended so as to cover such joint configurations:

- Mode 1 (thin flanges – full plastic mechanism in the flange):

$$
F_{T,Rd,1} = 2\pi m_{pl,Rd}\left[1 + \frac{R_1 + R_2}{m}\right]
\tag{3.37}
$$

- Mode 2 (intermediate flanges – yield lines in the flange + bolt failure):

$$
F_{T,Rd,2} = \frac{2\pi m_{pl,Rd} + n\Sigma F_{t,Rd}}{m + n}
\tag{3.38}
$$

- Mode 3 (thick flanges – bolt failure):

$$F_{T,Rd,3} = \sum F_{t,Rd}$$

(3.39)

These expressions are valid if

$$R_2 \leq N \, \frac{\min\left(2\pi m; 4m + 1.25n\right)}{2\pi}$$

(3.40)

In these formulae, the various parameters are defined according to Figure 3.36 and section 3.5.2.2; N is equal to the number of regularly spaced connecting bolts.

The failure mode of the connection is that corresponding to the minimum of three above defined design resistances.

UK Specific Comment

T-stub configurations with four bolts in a row are rarely used in the UK.

3.5.3 Gusset plates

Gusset plates are commonly used in steel construction. The connected members are either bolted or welded to the gusset plate. But in many cases, both types of connectors are selected, as shown in Figure 3.37.

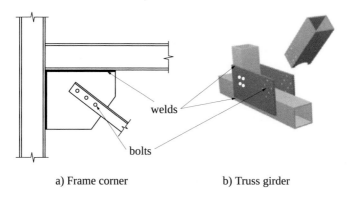

a) Frame corner   b) Truss girder

Figure 3.37 – Examples of gusset plate connections using welds and bolts

In the vicinity of the gusset, several failure modes may occur at different locations:

- In the supported element(s);
- In the bolted or welded connection(s) between the supported member(s) and the gusset plate;
- In the gusset plate;
- In the bolted or welded connection(s) between the gusset plate and the supported member(s);
- In the supported member(s).

In the present section, only the way on how the gusset plate resists to the applied forces is considered as all the other failure modes are covered elsewhere in chapters 3 and 4.

In fact, two categories of failure modes are likely to develop in the gusset plate, independently of the kind of connectors used (bolts or welds):

- Local failure in the gusset plate under the forces transferred by the supported member(s);
- Global cross section failure of the gusset plate under the resultant of forces transferred by more than one of the supported member(s).

These ones are illustrated in Figure 3.38 in the particular case of a gusset plate connecting two diagonals and a main chord in a truss girder.

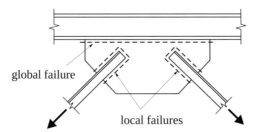

Figure 3.38 – Failure modes in a gusset plate

In EN 1993-1-8, no specific guidelines are provided for the local and global verifications of the gusset plate, or at least no full consistent set of design formulae. The following recommendations may however be proposed:

a) Local failure (supported member in tension)

A block tearing type failure has to be checked, both for welded and bolted connections, as illustrated in Figure 3.39. To achieve it, reference may be simply made to the corresponding EN 1993-1-8 verification formulae (see section 3.4.6).

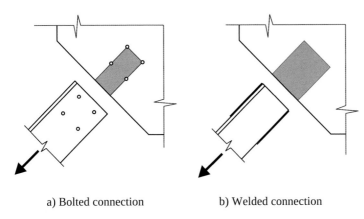

a) Bolted connection   b) Welded connection

Figure 3.39 – Block tearing failure

b) Local failure (supported member in tension)

For supported members in compression, gusset local plate instability may occur. In this context, good practice indicates that the region of unsupported gusset plate at the ends of the members should be kept to a minimum. End and edge distance requirements for the bolts should however be strictly followed.

No information is provided in EN 1993-1-8 for such a stability check. In Disque and Robert (1984), the authors recommend to allow for the local plate instability of the gusset in the flexural buckling of a column:

- With a buckling length defined as the average of l_1, l_2 and l_3 (see Figure 3.40) multiplied by a buckling length coefficient K assumed to be equal to 0.65, or even 0.5 by some authors (Gross, 1990);
- With a unit width;
- Subjected to compressive stresses equal to those defined on the Whitmore section (Whitmore, 1952), i.e. on an effective width defined as specified in Figure 3.41.

115

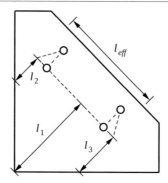

Figure 3.40 – Buckling lengths

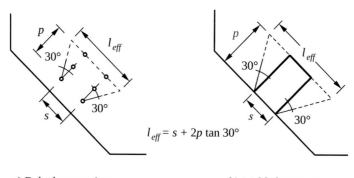

$$l_{eff} = s + 2p \tan 30°$$

a) Bolted connection   b) Welded connection

Figure 3.41 – Gusset plate yielding

A safe application of this concept, assuming that the Whitmore stress equals the yield strength of the material, leads to an easy-to-apply design criterion:

$$Kl \leq 5.4t\varepsilon \tag{3.41}$$

with:

$\varepsilon = \sqrt{\dfrac{235}{f_{y,g}}}$ coefficient dependent on the yield strength of the gusset

plate $f_{y,g}$ (expressed in N/mm^2)

c)   Global failure (supported member in tension)

Reference is made here to sub-chapter 1.6 where the "static approach" is introduced. The verification of the global resistance therefore comprises the checking the resistance of all potential cross sections in the gusset plates, see Figure 3.42. Elastic or plastic stress distributions may be considered under the applied forces.

Potential critical section subjected to:

- axial force $(F_{Ed,1} - F_{Ed,2})/\sqrt{2}$

- shear force $(F_{Ed,1} - F_{Ed,2})/\sqrt{2}$

- moment $(F_{Ed,1} + F_{Ed,2})d/2\sqrt{2}$

Figure 3.42 – Example of verification for the global failure mode

---

UK Specific Comment

Reference is made to the SCI method for the design of gusset plates, see SCI (2011).

3.5.4 Long joints

The evaluation of the design resistance of bolted lap joints using preloaded and non-preloaded bolts has been discussed in section 3.5.1 and design formulae have been provided.

In Category A (bearing type) and Category B (slip resistance at SLS) connections, a reduction factor has however possibly to be applied for so-called "long joints", i.e. joints with a significant distance between the two extreme bolt rows. In such long joints, and contrary to what has been shown for short joints in section 3.5.1, the deformation capacity in the end bolt rows is too limited to allow a full plastic redistribution of the internal forces to take place between the bolt rows, even if the design shear resistance $F_{v,Rd}$ of each individual fastener is greater than or equal to the local design bearing resistance $F_{b,Rd}$.

Practically, EN 1993-1-8 recommends to reduce the design resistance of the joint by a factor β_{Lf} where the distance L_j between the centres of the end fasteners in a joint, measured in the direction of force transfer (see Figure 3.43), is larger than $15d$, and β_{Lf} is defined as:

$$\beta_{Lf} = 1 - \frac{L_j - 15d}{200d} \tag{3.42}$$

but with $0.75 \leq \beta_{Lf} \leq 1.0$.

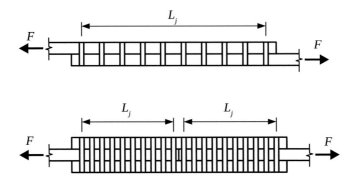

Figure 3.43 – Lap joint length

It should be noted that this reduction factor never applies where there is a uniform transfer of forces over the length of the joint, e.g. the transfer of longitudinal shear forces between the web and the flange of a section in a beam in bending.

Chapter 4

WELDED CONNECTIONS

4.1 TYPES OF WELDS

Fillet welds and butt welds are widely used in construction. They represent 80% and 15%, respectively, of all welds found in structures. For the remaining five per cent, plug and fillet all round welds are mostly used. In the following paragraphs, these different weld types are briefly introduced.

4.1.1 Butt welds

Butt welds are used between adjacent plates in so-called butt and tee-joints (Figure 4.1a). The preparation of the plates, before welding, usually appears as a prerequisite, except for thin plates, less than about 5 mm, where it can be avoided. In all the other cases, bevelled plate edges will have to be prepared, see Figures 4.1b and 4.2. This may have an impact on the economy of the project.

119

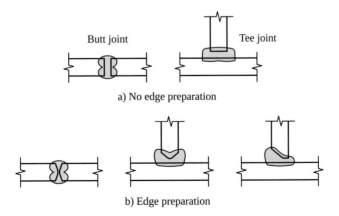

Figure 4.1 – Butt welds with full penetration

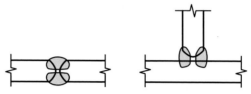

Figure 4.2 – Examples of types of bevelled edges

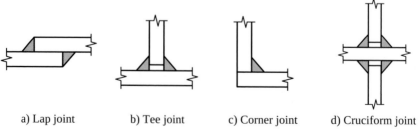

Figure 4.3 – Butt welds with partial penetration

According to the penetration of the weld, i.e. the degree of fusion of the weld and parent materials through the thickness of the connected plates, butt welds with complete penetration (equal to the plate thickness, see Figure 4.1) or partial penetration (less than the plate thickness, see Figure 4.3) are distinguished.

4.1.2 Fillet welds

Contrary to a butt weld, a fillet weld requires no preparation of the plates to connect. This is an important feature in terms of fabrication costs.

Fillet welds are applied to the surface of the connected plates; they exhibit an approximately triangular shape.

Lap joints, tee or cruciform joints and corner joints are shown in Figure 4.4. In tee joints, the two plates are not necessarily welded perpendicularly each to another.

a) Lap joint   b) Tee joint   c) Corner joint   d) Cruciform joint

Figure 4.4 – Schematic representation of various fillet weld joint configurations

As explained later in section 4.4.2.4, the one-sided character of the fillet weld in the corner joint represented in Figure 4.4c is likely to negatively influence the joint response. As a way of improvement, the outer corner could be butt welded (Figure 4.5).

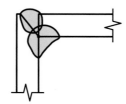

Figure 4.5 – Improved corner joint

Fillet welds are either continuous or intermittent along the whole joint, Figure 4.6. The latter should not be used in corrosive environments.

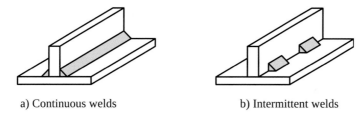

a) Continuous welds   b) Intermittent welds

Figure 4.6 – Continuous and intermittent fillet welds

4.1.3 Fillet welds all round

Fillet welds all round is a term used to describe fillet welds in circular or elongated holes, see Figure 4.7. Such welds may be used only to transmit shear or to prevent the buckling or separation of lapped parts.

Figure 4.7 – Fillet welds all round

4.1.4 Plug welds

As seen in Figure 4.8, the holes (round or elongated) formed in one of the two plates to assemble are filled with welding material.

Such welds may be used:

– To transmit shear;
– To prevent the buckling or separation of lapped parts;
– To inter-connect the components of built-up members;

but should not be used to resist externally applied tension.

a) Round holes   b) Elongated holes

Figure 4.8 – Plug welds

4.2   CONSTRUCTION CONSTRAINTS

Welding is a quite convenient assembly technique, but its reliability is highly dependent on the respect of strict requirements in terms of properties of parent and consumable materials, welding process, tolerances, preparation of surfaces, storage and handling of consumables, weather protection, welding sequence, preheating, operator qualification, acceptance criteria, etc. These aspects are extensively covered in EN 1090-2, and require from the manufacturers high competence on the engineering and operating sides.

This chapter focuses on design aspects (sub-chapter 4.3) and therefore these technological aspects will simply be briefly addressed below.

4.2.1 Mechanical properties of materials

4.2.1.1 Parent material

The design provisions proposed in sub-chapter 4.3 apply to weldable structural steels conforming to EN 1993-1-1 and to material thickness values of 4 mm and over. The provisions also apply to joints in which the mechanical properties of the weld metal are compatible with those of the parent metal, see section 4.2.1.2.

For welds in thinner material reference should be made to EN 1993-1-3 (CEN, 2006f) and, for welds in structural hollow sections in material thickness of 2.5 mm and over, guidance is given in EN 1993-1-8.

Quality level C according to EN ISO 5817 (CEN, 2014) is usually required, if not otherwise specified.

The choice of material as well as the verification for lamellar tearing should be performed according to EN 1993-1-10 (CEN, 2005e).

4.2.1.2 Welding consumables

All welding consumables should conform to the relevant standards specified in EN 1090-2. The specified yield strength, ultimate tensile strength, elongation at failure and minimum Charpy V-notch energy value of the filler metal, should be equivalent to, or better than that specified for the parent material.

Generally it is safe to use electrodes that are overmatched with regard to plates in steel grades up to S460. For higher steel grades up to S700 undermatching needs to be accounted for according to the rules given in EN 1993-1-12 (CEN, 2007a).

UK Specific Comment

It is anticipated that specific rules given in the current BS EN 1993-1-12 (BSI, 2009) for high strength steels will be incorporated in the revised BS EN 1993-1-8, to appear after 2020.

4.2.2 Welding processes, preparation of welds and weld quality

One of the following arc welding processes may be used:

– Metal arc-welding with covered electrodes;
– Flux-cored arc welding;

- Submerged arc welding;
- Metal Inert Gas (MIG) welding;
- Metal Active Gas (MAG) welding.

All these processes can be used in the workshop. Usually, only bolting or metal arc welding with covered electrodes is employed in the erection phase on the building site. With the metal arc welding process, welds can be made in all positions. The various weld positions are shown in Figure 4.9, where the arrows give the arc direction during the welding operation. It is clear that welding in the flat position is easily carried out, allowing a greater rate of metal deposition than the other positions; by welding in this favourable position, the maximum size of weld run can be obtained. With ordinary welding consumables and favourable welding conditions, a fillet weld with a throat thickness of 6 mm can be produced with only one run. For welds of greater thickness, more than one run is necessary. In this case, the welding sequence must be carefully planned, see Figure 4.10.

The welding conditions, particularly the current limitation of the welding equipment, constitute a limit to the depth of penetration into the parent plate. For example, if a closed butt joint (no gap between the two plates) is welded with one run on each side, the penetration may not be complete and the central part of the joint will remain unfused (Figure 4.11a). With a gap between the two parts of the joint, full penetration can be achieved with the same welding equipment. The limitation is then set by the thickness of the plates to be joined. In practice, the limit for butt welds with square edges, i.e. without preparation, is 10 mm plate thickness with a 5 mm gap. When the plate thickness exceeds this value, bevelled edges permit full penetration by several runs, see Figure 4.11b.

Edge preparation consists essentially of cutting and bevelling the edges of the plates which are to be welded. These operations can be done by thermal cutting, by machining or by chipping or grinding. The resulting surfaces of the bevelled edges should be smooth, uniform, free from cracks and without corrosion. If thermal cutting or another process which hardens the material is used, the approved welding procedure must take account of this weld preparation process. Different bevel geometries are shown in Figure 4.2.

a) Flat position

b) Horizontal-vertical position

c) Vertical position

d) Overhead position

Figure 4.9 – Weld positions

Figure 4.10 – Welds with successive runs

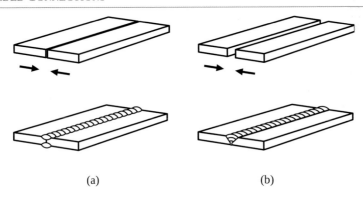

(a)   (b)

Figure 4.11 – Effects of the gap on weld penetration

Practical recommendations for tolerance values for various weld types are given, for example, in EN 1993-1-9 for fatigue (CEN, 2005d) or EN 1993-2 for steel bridges (CEN, 2006e).

Before welding, surfaces and edges adjacent to the weld location must be cleaned to remove oil, grease, paint or any other contaminants, which can affect the quality of the weld and the weld strength.

The appropriate welding method and procedure are defined in a project specification provided by the designer. Relative information may be found in EN 1090-2.

Quality control is an important part of industrial activity. The term quality includes all the characteristics of a product which affect its ability to serve its purpose. Here, attention is drawn to quality control applied to welding, including the qualification of firms and the procedure qualification for welding tests.

The principal purpose of weld inspection is to find possible weld defects. Examples of weld defects are, see Figure 4.12:

- *Undercut*: the thickness of the parent metal is reduced near the weld toe;
- *Porosity or gas inclusions*: air or gas bubbles are incorporated in the melted metal, where they remain after cooling;
- *Insufficient throat*: the throat thickness is smaller than the design thickness. The resistance of the joint might be insufficient;
- *Incomplete penetration*: The throat thickness is smaller than the design thickness. The resistance of the joint might be insufficient.

All these defects can be measured. Codes of practice specify the allowable tolerances for each defect, see EN ISO 5817 (CEN, 2014).

a) Undercut   b) Porosity or gas inclusions

c) Incomplete penetrations

Figure 4.12 – Examples of weld defects

4.2.3 Geometry and dimensions of welds

4.2.3.1 Fillet welds

In tee-joints, fillet welds may be used to connect parts where the fusion faces form an angle between 60° and 120°. Angles smaller than 60° are also permitted; however, in such cases, the weld should be considered to be a partial penetration butt weld. For angles greater than 120° the resistance of fillet welds should be determined by testing in accordance with EN 1990 Annex D: Design by testing (CEN, 2002).

Fillet welds finishing at the ends or sides of parts should be returned continuously, full size, around the corner for a distance of at least twice the leg length of the weld, unless access or the configuration of the joint renders this impracticable. In the case of intermittent welds this rule applies only to the last intermittent fillet weld at corners. End returns should be indicated on the drawings.

4.2.3.2 Intermittent fillet welds

As already said, intermittent fillet welds should not be used in corrosive environments.

The gaps L_1 or L_2 between the ends of each length of weld L_w should fulfil the requirement given in Figure 4.13. They should be taken as the smaller of the distances between the ends of the welds on opposite sides and the distance between the ends of the welds on the same side.

In any run of intermittent fillet weld there should always be a length of weld at each end of the part connected.

In a built-up member in which plates are connected by means of intermittent fillet welds, a continuous fillet weld should be provided on each side of the plate for a length at each end equal to at least three-quarters of the width of the narrower plate concerned.

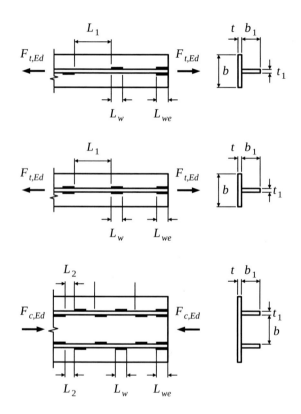

Requirements for the gaps (L_1, L_2):   min ($L_{we} \geq 0.75b$; $0.75b_1$)

For build-up members in tension: min ($L_1 \leq 16t$; $16\,t_1$; 200 mm)

For build-up members in compression or shear:

          min ($L_2 \leq 12t$; $12\,t_1$; $0.25b$; 200 mm)

Figure 4.13 – Geometry of intermittent fillet welds

4.2.3.3 Fillet welds all round

As specified earlier, fillet welds all round may be used only to transmit shear or to prevent the buckling or separation of lapped parts.

The diameter of a circular hole, or width of an elongated hole, for a fillet weld all round should not be less than four times the thickness of the part containing it.

The ends of elongated holes should be semi-circular, except for those ends which extend to the edge of the part concerned.

The centre-to-centre spacing of fillet welds all round should not exceed the value necessary to prevent local buckling, see Table 3.3 in sub-chapter 3.3.

4.2.3.4 Butt welds

Full penetration and partial penetration welds may be considered, as defined in section 4.3.4. Intermittent butt welds should not be used.

4.2.3.5 Plug welds

It has been said before that plug welds can be used:

- To transmit shear;
- To prevent the buckling or separation of lapped parts;
- To inter-connect the components of built-up members;

but should not be used to resist externally applied tension.

The diameter of a circular hole, or width of an elongated hole, for a plug weld should be at least 8 mm more than the thickness of the part containing it. The ends of elongated holes should either be semi-circular or else should have corners which are rounded to a radius of not less than the thickness of the part containing the slot, except for those ends which extend to the edge of the part concerned.

The thickness of a plug weld in parent material up to 16 mm thick should be equal to the thickness of the parent material. The thickness of a plug weld in parent material over 16 mm thick should be at least half the thickness of the parent material and not less than 16 mm.

As for fillet welds all round, the centre to centre spacing of fillet welds all round should not exceed the value necessary to prevent local buckling, see Table 3.3 in sub-chapter 3.3.

4.2.3.6 Welding in cold-formed zones

Welding may be carried out within a length 5*t* either side of a cold-formed zone (see Table 4.1), provided that one of the following conditions is fulfilled:

– The cold-formed zones are normalised after cold-forming but before welding;
– The *r*/*t*-ratio satisfy the relevant value obtained from Table 4.1.

Table 4.1 – Conditions for welding cold-formed zones and adjacent material

| *r*/*t* | Strain due to cold forming (%) | Maximum thickness (mm) | | Fully killed Aluminium-killed steel (Al ≥ 0.02%) |
| | | Generally | | |
		Predominantly static loading	Where fatigue predominates	
≥ 25	≤ 2	any	any	any
≥ 10	≤ 5	any	16	any
≥ 3.0	≤ 14	24	12	24
≥ 2.0	≤ 20	12	10	12
≥ 1.5	≤ 25	8	8	10
≥ 1.0	≤ 33	4	4	6

4.3 DESIGN OF WELDS

4.3.1 Generalities

For weld design, three fundamental assumptions are usually made:

– The welds are homogeneous and isotropic;
– The connected elements are rigid and their deformations are negligible;
– Only nominal stresses due to external loads are considered. Effects of residual stresses, stress concentrations and shape of the welds are neglected in static design.

These assumptions will be briefly discussed in sub-chapter 4.4.

4.3.2 Fillet welds

A fillet weld is characterised by two main geometrical properties, effective length and effective throat thickness, defined in sections 4.3.2.1 and 4.3.2.2. The design resistance of a fillet weld is derived from these two values, and the weld and base material grades, see section 4.3.2.3.

4.3.2.1 Effective length

The actual physical weld length L_w is usually substituted for an effective length L_{eff} for design purposes. L_{eff} is taken as the length over which the fillet is full-size, which is defined as the overall length of the weld reduced by twice the effective throat thickness a (see 4.3.2.2). If the weld is full size throughout its length, including starts and terminations, no reduction in effective length needs to be accounted for.

A weld with a length smaller than 30 mm or less than six times its throat thickness, whichever is larger, should not be designed to carry loads.

4.3.2.2 Effective throat thickness

The effective throat thickness of a fillet weld should be taken as the height of the largest triangle (with equal or unequal legs) that can be inscribed within the fusion faces and the weld surface, measured perpendicular to the outer side of this triangle (see Figure 4.14).

In practice, the effective throat thickness of a fillet weld should not be less than 3 mm.

In the case of a deep penetration fillet weld, a higher value for the throat thickness may be assumed, provided that preliminary tests show that the required penetration can consistently be achieved by the welder, see Figure 4.15.

131

Figure 4.14 – Throat thickness of a fillet weld

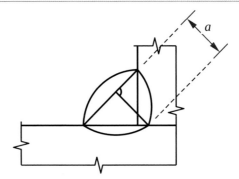

Figure 4.15 – Throat thickness of a deep penetration fillet weld

4.3.2.3 Design resistance

Two procedures for the evaluation of the design resistance of a fillet weld are provided in EN 1993-1-8, the directional and simplified methods.

The most economical one, i.e. the one providing greath strength for material used, is the directional method. In the case of longitudinal (or side) welds (i.e. weld parallel to the applied force) subjected to shear forces along the weld axis, equal results are obtained from both approaches.

4.3.2.3.1 Directional method

In this method, the forces transmitted by a unit length of weld are resolved into four components. These ones are respectively parallel and transverse to the longitudinal axis of the weld and normal and transverse to the plane of the throat:

$\sigma_\perp$ = normal stress perpendicular to the throat,

$\sigma_{\|}$ = normal stress parallel to the axis of the weld,

$\tau_\perp$ = shear stress (in the plane of the throat) perpendicular to the axis of the weld,

$\tau_{\|}$ = shear stress (in the plane of the throat) parallel to the axis of the weld.

These stresses, shown in Figure 4.16, are assumed to be uniformly distributed on the throat section $A_w = L_{eff}\, a$.

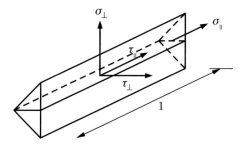

Figure 4.16 – Stresses on the throat section of a fillet weld (unit length)

The design resistance of the fillet weld will be sufficient if the following equations are both satisfied:

$$\sqrt{\left[\sigma_{\perp}^2 + 3\left(\tau_{\perp}^2 + \tau_{\parallel}^2\right)\right]} \le \frac{f_u}{\beta_w \, \gamma_{M2}} \quad \text{and} \quad \sigma_{\perp} \le \frac{0.9 f_u}{\gamma_{M2}} \tag{4.1}$$

where:

f_u =   nominal ultimate tensile strength of the weaker part joined;

β_w =   appropriate correlation factor taken from Table 4.2.

The value of $\sigma_{\parallel}$ does not influence the resistance of the weld.

When the connected members exhibit different material strength grades, the design of the weld should be based on the mechanical properties of the material with the lower strength grade.

133

Table 4.2 – Correlation factor β_w for fillet wed design

Standard and steel grade			Correlation factor
EN 10025	EN 10210	EN 10219	β_w
S 235 S 235 W	S 235 H	S 235 H	0.8
S 275 S 275 N/NL S 275 M/ML	S 275 H S 275 NH/NLH	S 275 H S 275 NH/NLH S 275 MH/MLH	0.85
S 355 S 355 N/NL S 355 M/ML S 355 W	S 355 H S 355 NH/NLH	S 355 H S 355 NH/NLH S 355 MH/MLH	0.9

Table 4.2 – Correlation factor β_w for fillet wed design (continuation)

Standard and steel grade			Correlation factor
EN 10025	EN 10210	EN 10219	β_w
S 420 N/NL S 420 M/ML		S 420 MH/MLH	1.0
S 460 N/NL S 460 M/ML S 460 Q/QL/QL1	S 460 NH/NLH	S 460 NH/NLH S 460 MH/MLH	1.0

4.3.2.3.2 Simplified method

As an alternative to section 4.3.2.3.1, the design resistance of a fillet weld may be assumed to be adequate if, at every point along its length, the resultant of all the forces per unit length transmitted by the weld satisfy the following criterion:

$$F_{w,Ed} \leq F_{w,Rd} \tag{4.2}$$

where:

$F_{w,Ed}$ = design value of the weld force per unit length;

$F_{w,Rd}$ = design weld resistance per unit length.

The resistance of the weld per unit length $F_{w,Rd}$ is defined independently of the orientation of the weld throat plane relatively to the applied force, as follows:

$$F_{w,Rd} = f_{vw,d}\, a \tag{4.3}$$

where the design shear strength of the weld $f_{vw,d}$ is defined as:

$$f_{vw,d} = \frac{f_u}{\sqrt{3}\beta_w\, \gamma_{M2}} \tag{4.4}$$

— checked in BS

$\gamma_{M2} = 1\text{-}25$

The values for f_u and β_w are defined in section 4.3.2.3.1.

4.3.3 Fillet welds all round

The design resistance of a fillet weld all round is evaluated by using the same expressions than those presented in section 4.3.2.

4.3.4 Butt welds

Providing that the welding process has been correctly carried out, the butt weld filler metal may be considered as parent metal. Hence, to determine the resistance of the joint, the calculation is based on the throat area, i.e. the penetration area. Depending on the penetration, two types of butt welds are defined: full and partial penetration welds.

For a full penetration butt weld, usually in overmatching situations, calculation is not necessary because the filler metal strength is at least as high as the parent metal strength of the weaker connected element and the throat thickness of the weld is equal to the thickness of the plate, see Figure 4.17. Thus the butt weld may effectively be regarded simply as replacing the parent material.

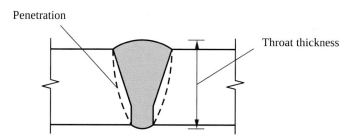

Figure 4.17 – Butt weld with full penetration

The evaluation of the design resistance of a partial penetration butt weld, see Figure 4.18, is similar to the one prescribed for a deep penetration fillet weld. The throat thickness of a partial penetration butt weld should not be greater than the depth of penetration that can be consistently achieved.

135

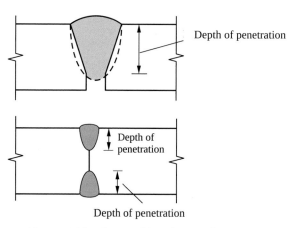

Figure 4.18 – Butt weld with partial penetration

A partial penetration tee-butt joint with superimposed fillet welds may be considered as a full penetration butt weld, if the total throat thickness is greater than the material thickness and the gap dimension meets certain conditions, see Figure 4.19. If it is not the case, the resistance should be determined using the method for a fillet weld or a deep penetration fillet weld, depending on the amount of penetration. The throat thickness should be determined in conformity with the provisions for fillet welds or partial penetration butt welds, as relevant.

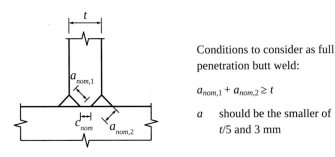

Conditions to consider as full penetration butt weld:

$a_{nom,1} + a_{nom,2} \geq t$

a should be the smaller of $t/5$ and 3 mm

Figure 4.19 – Tee-butt joint with superimposed fillet welds

4.3.5 Plug welds

The design resistance $F_{w,Rd}$ of a plug weld is given by:

$$F_{w,Rd} = f_{vw,d} A_w \qquad (4.5)$$

where:

$f_{vw,d}$ is   the design shear strength of a weld given by Eq. (4.4);
A_w is   the design throat area defined as the area of the hole.

4.3.6 Concept of full strength fillet weld

Fillet welds must be designed according to sections 4.3.2.3.1 or 4.3.2.3.2. In the case of relatively small loads in relation to the capacity of the weld, EN 1993-1-8 design rules may lead to rather small welds. If the resistance of those thin welds is lower than the yield strength of the weakest of the connected parts, the connection has so little deformation capacity that it is usually not sufficient to accommodate effects due to imposed deformations, for instance. In such a case the connection will behave in a brittle way.

To avoid this, the welds should be designed "full strength". The resistance of full strength welds would then be greater than the resistance of the connected elements; so, in the case of overloading, the elements would fail before the welds.

A similar requirement is also regularly imposed on connections where global ductility is required. This aspect will be addressed in some different further chapters of the present book, as sub-chapter 4.4 for welded joints, chapter 5 for simple joints, and chapter 8 for joints under various "other than static" loading situations.

The "full strength" character of the welds may be achieved by ensuring that the design resistance of the weld is higher than or equal to the design resistance of the weakest connected element. As an example, the derivation of "full strength" throat thicknesse is illustrated here below in the case of double fillet end welds (i.e. weld perpendicular to the applied force). If reference is made to the directional method, the design resistance of a fillet weld is checked as follows:

$$\sigma_c = \sqrt{\sigma_\perp^2 + 3\tau_\perp^2 + 3\tau_{//}^2} \le \frac{f_u}{\beta_w \, \gamma_{M2}} \quad \text{and} \quad \sigma_\perp \le 0.9 \frac{f_u}{\gamma_{M2}} \qquad (4.6)$$

where:

f_u =   nominal ultimate tensile strength of the weaker part joined

γ_{M2} =   partial safety factor for welded connections

β_w =   correlation factor (β_w = 1.0 for steel grades S420 and S460, see Table 4.3)

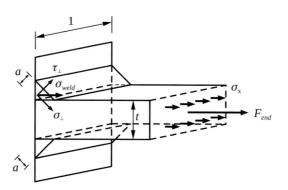

Figure 4.20 – End fillet and side fillet welds

For end fillet welds (Figure 4.20):

$$\sigma_\perp = \tau_\perp = \frac{\sigma_{weld}}{\sqrt{2}} \tag{4.7}$$

$$\tau_\perp = 0 \tag{4.8}$$

From Eqs. (4.6), (4.7) and (4.8) (more restrictive here than the second one), it follows:

$$\sigma_c = \sqrt{\left(\frac{\sigma_{weld}}{\sqrt{2}}\right)^2 + 3\left(\frac{\sigma_{weld}}{\sqrt{2}}\right)^2} \leq \frac{f_u}{\beta_w \gamma_{M2}} \tag{4.9}$$

and so:

$$\sigma_{weld} \leq f_{w,u,end} = \frac{f_u}{\beta_w \gamma_{M2} \sqrt{2}} \tag{4.10}$$

For double end (transverse) fillet welds, the above-expressed "full strength" criterion writes (Figure 4.20):

$$2 a L_w \sigma_{weld} \geq \frac{t L_w f_y}{\gamma_{M0}} \tag{4.11}$$

This last equation is obtained by assuming, in Figure 4.20, that $F_{end} = \sigma_x t l_w$ with $\sigma_x = f_y/\gamma_{M0}$.

The minimum weld size to satisfy the full strength design requirement is therefore expressed as:

$$a \geq \frac{f_y t}{2 f_{w,u,end} \gamma_{M0}} \tag{4.12}$$

and finally:

$$a \geq \frac{f_y \beta_w \gamma_{M2}}{\sqrt{2} f_u \gamma_{M0}} t \tag{4.13}$$

Minimum throat thickness values required for a full strength end weld design are reported in Table 4.3 for various steel grades.

Table 4.3 – Values of β_w and $f_{w,u,end}$ for steels according to EN 10025 and minimum "full strength" required weld thickness in case of double fillet end welds (plate thickness smaller than 40 mm; $\gamma_{M0} = 1.0$ and $\gamma_{M2} = 1.25$)

Steel grade	f_y	f_u	β_w	$f_{w,u,end}$	Full strength double fillet welds
	N/mm^2	N/mm^2		N/mm^2	
S235	235	360	0.80	255	$a \geq 0.46\,t$
S275	275	430	0.85	286	$a \geq 0.48\,t$
S355	355	510	0.90	321	$a \geq 0.55\,t$
S420 M	420	520	1.00	294	$a \geq 0.71\,t$
S420 N	420	550	1.00	311	$a \geq 0.68\,t$
S460 M	460	550	1.00	311	$a \geq 0.74\,t$
S460 N	460	580	1.00	328	$a \geq 0.70\,t$

4.4 DISTRIBUTION OF FORCES IN A WELDED JOINT

4.4.1 Generalities

This topic has been addressed in chapter 1 (see section 1.6.1) in which reference is made to the so-called "static approach". For welded joints, however, the limited ductility of the welds may lead to a limitation of its application even if, according to EN 1993-1-8, "the distribution of forces in a welded connection may be calculated on the assumption of either elastic or plastic behaviour". In reality:

- Welds are characterised by the presence of residual stresses as well as of geometrical imperfections; the latter in local stress concentrations. So even before a loading is applied, variations of internal stresses and strains are present along the weld. Locally, the residual stresses could even be higher than the material strength and so provoke failure; but thanks to the small amount of available ductility of the weld material, plastic redistribution takes place and the stresses are just limited locally to the material yield strength;

- When the welds are then externally loaded, new plastic redistributions have to take place to allow, at the end, to reach a uniform distribution of stresses equal to the yield strength along the design throat area. To reach such a plastic distribution of internal stresses, further local ductility is obviously required.

In section 4.3.1, it has been assumed that the effect of these various imperfections may be neglected for static design. This does not mean that the ductility of the weld is always sufficient to rely on a full plastic redistribution of internal forces along the weld length. EN 1993-1-8 does not explicitly define what is acceptable, even if some clauses may be interpreted as a way of limiting plastic redistributions.

In the paragraphs below, an attempt is made to clarify the point.

When a weld is subjected to pure shear forces, or axial forces, the ductility of the weld is assumed to be sufficient to reach a full plastic redistribution of stresses along the weld, but only as long as some geometrical requirements or layouts are strictly fulfilled (see section 4.4.2.1).

In other cases, as a general statement, it is recommended to adopt an elastic distribution of forces as long as partial strength welds are used. Doing so reduces the risk of brittle failure resulting from a local yielding of a weld segment. Consequently, a plastic distribution of forces is seen to be basically limited to cases where full strength welds are adopted (the local ductility being so provided by the parent material). In Figure 4.21, both situations are illustrated. Obviously, some exceptions may be envisaged because the plastic redistribution of forces only requires a quite limited deformation capacity of the weld, see Figure 4.22.

Besides that, it is worth recalling that residual stresses and stresses not subjected to transfer of load (as long as they remain rather limited) do not need to be included when checking the resistance of a weld. This applies specifically to the normal stresses parallel to the axis of a weld.

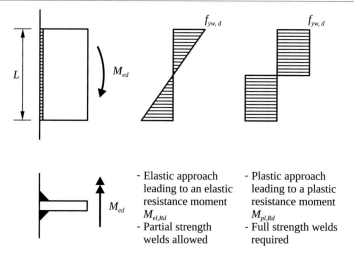

- Elastic approach leading an elastic resistance moment $M_{el,Rd}$
- Partial strength welds allowed

- Plastic approach leading to a plastic resistance moment $M_{pl,Rd}$
- Full strength welds required

Figure 4.21 – Elastic and plastic distributions of forces

- Let assume $F_{Rd,side} < F_{Rd,end}$
- Let assume equal rigidities for end and side welds

  - $F_{el,Rd} = 3\ F_{Rd,\ side}$
  - $F_{pl,Rd} = 2\ F_{Rd,side} + F_{Rd,end}$

- $F_{pl,Rd}$ accepted by accounting for a relative ductility of side welds compared to end weld

Figure 4.22 – Example of elastic and plastic distributions

As EN 1993-1-8 does not provide detailed recommendations on this matter, a conservative, but pragmatic, approach is expected from the designer. In section 6.3.5, a particular application of the static theorem to beam-to-end plate connections will be presented.

4.4.2 Particular situations

4.4.2.1 Long lap joints with side fillet welds

In lap joints, the distribution of stresses in the side fillet welds is usually assumed to be uniformly distributed along the weld length (see Figure 4.23a) and the design resistance is evaluated accordingly. The actual distribution may however differ significantly from this assumption, as larger stresses occur at the ends of the connection, see Figure 4.23b. This has similarly been noted in section 3.5.4 for long bolted joints. At ULS, just before failure, plastic deformation near the ends may contribute to a more uniform shear stress in the welds. However, in long connections, the limited available ductility in the welds resulting from local plasticity does not allow for a full plastic redistribution and the resulting stress distribution is not fully uniform.

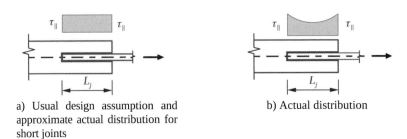

a) Usual design assumption and approximate actual distribution for short joints

b) Actual distribution

Figure 4.23 – Elastic distribution of shear stresses along the welds

EN 1993-1-8 specifies that the design resistance of welded lap joints can be evaluated on the basis of an assumed uniform distribution of shear stresses for joints with a length lower or equal to $150a$. For longer joints, the evaluated resistance should be multiplied by a reduction factor $\beta_{L,w}$ ($\beta_{L,w,1}$ or $\beta_{L,w,2}$) to allow for the effects of non-uniform stress distribution:

a)   Lap joints longer than $150a$:

$$\beta_{Lw,1} = 1.2 - \frac{0.2L_j}{150a} \leq 1.0 \qquad (4.14)$$

L_j = overall length of the lap in the direction of the force transfer, see Figure 4.23.

b)   For fillet welds longer than 1.7 m connecting transverse stiffeners in plated members

$$\beta_{Lw,2} = 1.1 - \frac{L_w}{17} \quad \text{but } 0.6 \leq \beta_{Lw,2} \leq 1.0 \qquad (4.15)$$

L_w = length of the weld (in metres)

4.4.2.2 Welds to unstiffened flanges

If a plate is welded to an unstiffened flange of an I, H or a box section, loading will tend to deform the flange or the box side unequally along the breadth. The result is that the parts of the weld near the web will be more heavily loaded than the other parts, see Figure 4.24. Therefore, a reduced effective width has to be taken into account for both the parent material and the weld design.

Figure 4.24 – Effective width for unstiffened tee-joints

a)   For an I or H section the effective width b_{eff} should be taken as:

$$b_{eff} = t_w + 2s + 7kt_f \qquad (4.16)$$

where:

$$k = \left(\frac{t_f}{t_p}\right)\left(\frac{f_{y,f}}{f_{y,p}}\right) \quad \text{but} \leq 1.0 \qquad (4.17)$$

$f_{y,f}$ =   design yield strength of the flange of the I or H section
$f_{y,p}$ =   design yield strength of the welded plate

The geometrical parameters t_w, t_f and t_p are shown in Figure 4.24 and s is:

- For a rolled I or H section: $s = r$ (root radius)
- For a welded I or H section: $s = \sqrt{2}a$

However, for the unstiffened flange of an I or H section, the following criterion should be satisfied:

$$b_{eff} \geq \left(\frac{f_{y,p}}{f_{u,p}} \right) b_p \qquad (4.18)$$

where:

$f_{u,p}$ = ultimate strength of the plate welded to the I or H section;
b_p = width of the plate welded to the I or H section.

If this requirement is not satisfied then the supporting member should be stiffened.

b) For other sections such as box sections or channel sections where the width of the connected plate is similar to the width of the flange, the effective width b_{eff} should be obtained from:

$$b_{eff} = 2t_w + 5t_f \quad \text{but} \quad b_{eff} \leq 2t_w + 5kt_f \qquad (4.19)$$

4.4.2.3 Intermittent welds

As an extra rule, EN 1993-1-8 indicates that if the design resistance of an intermittent weld is determined by using the total length l_{tot}, the weld shear force per unit length $F_{w,Ed}$ should be multiplied by the factor $(e + l)/l$ (see Figure 4.25).

Figure 4.25 – Geometry of intermittent welds

4.4.2.4 Eccentrically loaded single fillet or single-sided partial penetration butt welds

Local eccentricity should be avoided whenever it is possible through appropriate conceptual design. When present, local eccentricities (relative to the line of action of the force to be resisted) should be taken into account in the following cases:

- Where a bending moment transmitted about the longitudinal axis of the weld produces tension at the root of the weld (Figure 4.26a);
- Where a tensile force transmitted perpendicular to the longitudinal axis of the weld produces a bending moment, resulting in a tension force at the root of the weld (Figure 4.26b).

Obviously, local eccentricity has not to be taken into account if a weld is used as part of a weld group around the perimeter of a structural hollow section.

a) Bending moment produces   b) Tension force produces tension
tension at the root of the weld   at the root of the weld

Figure 4.26 – Single fillet or single-sided partial penetration butt welds

145

4.4.2.5 Angles connected by one leg

In section 3.4.11, attention has been drawn to the presence of eccentricities in angles connected by bolts located on one leg only, for instance to a gusset plate. The same occurs for similar welded joints. Here again a simplified solution may be found by adopting an effective cross sectional area and then treating the member as concentrically loaded.

For an equal-leg angle, or an unequal-leg angle connected by its larger leg, the effective area may be taken as equal to the gross area.

For an unequal-leg angle connected by its smaller leg, the effective area should be taken as equal to the gross cross sectional area of an equivalent equal-leg angle of leg size equal to that of the smaller leg, when determining the design resistance of the cross section. However, when determining the design buckling resistance of a compression member, the actual gross cross sectional area should be used.

4.4.2.6 Gusset plates

When gusset plates are used in a joint, they may be either connected by bolts or welds to each of the connected members. For the design of the weld, no further aspects need to be addressed here while, for the gusset plate, design rules have already been proposed in section 3.5.3.

Chapter 5

SIMPLE JOINTS

5.1 INTRODUCTION

Simple joints are assumed to transfer no bending moments and should be designed accordingly. They are only intended to resist axial and shear forces. In the following figures, various joint configurations where simple joints are likely to be used are illustrated. They are grouped in five categories:

- Beam-to-column joint configurations (Figure 5.1);
- Beam-to-beam joint configurations (Figure 5.2);
- Beam splice joint configurations (Figure 5.3);
- Bracing joint configurations (Figure 5.4);
- Column base joint configurations (Figure 5.5).

The connected members may be either steel or composite. Design rules for beam-to-column, beam-to-beam configurations and column bases are provided in this chapter. As the principles may be readily applied to splices, such joints are not explicitly addressed. The same applies to connections of bracings, the design of which is covered by rules already provided in chapters 3 and 4.

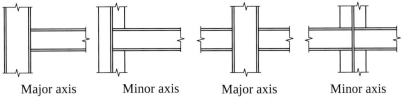

| Major axis | Minor axis | Major axis | Minor axis |

a) Single-sided joint configurations   b) Double-sided joint configurations

Figure 5.1 – Beam-to-column joint configurations

Un-notched   Single notched   Double notched

b) Single-sided joint configurations

Un-notched   Single notched   Double notched

b) Double-sided joint configurations

Figure 5.2 – Beam-to-beam joint configurations

Figure 5.3 – Beam splices and possible locations of simple beam splice joints

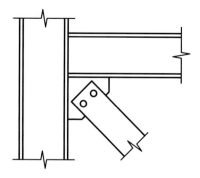

Figure 5.4 – Bracing configuration

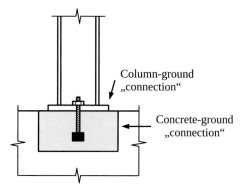

Figure 5.5 – Column base joint configuration

5.2 BEAM-TO-COLUMN AND BEAM-TO-BEAM JOINTS

5.2.1 Introduction

EN 1993-1-8 does not give specific design rules for simple joints. Formulae are provided for some joint components, but no full consistent set of design rules is made available. On the other hand, in some European countries, national rules for simple structural joints have existed for many years. Unfortunately, such recommendations did not cover all the possible types of failure and sometimes gave significantly different design rules for a typical failure mode.

As a first step, a comparative study of such available design rules for simple connections has been performed (Guillaume, 2000). In this work, reference has been made to different normative documents or design recommendations:

- EN 1993-1-1 (CEN, 2005a) and EN 1993-1-8 (CEN, 2005c);
- BS 5950 (BSI, 2000) and BCSA-SCI recommendations (BCSA/SCI Connections Group, 1993, 1992);
- NEN 6770 (NEN, 1997; Dutch Commissie SG/TC-10a (Verbindingen), 1998);
- German "Ringbuch" (Sedlacek *et al*, 2000).

Each of these documents covers its own application field, in which a limited number of possible failure modes will potentially occur. So, the comparison between them was difficult.

149

With the aim of establishing a full design approach according to the general design principles stated in EN 1993-1-8, design sheets for header plate and fin plate connections were prepared at the University of Liège and discussed at several meetings of Technical Committee 10 "Connections" of the European Convention for Constructional Steelwork (ECCS). This section presents these design rules, as finally agreed by ECCS in the form of "European Design Recommendations for the Design of Simple Joints in Steel Structures" (Jaspart *et al*, 2009).

The practical design recommendations presented in the afore-mentioned publication could possibly one day be implemented in EN 1993-1-8, as it is already the case in the last published version of the BCSA-SCI recommendations (Moreno *et al*, 2011) and the German "Ringbuch" (Weynand *et al*, 2013).

5.2.2 Scope and field of application

5.2.2.1 Types of structure

Simple structural joints are commonly found in steel framed buildings but they can also be used in other types of structures to connect steel elements (for example in bridges).

5.2.2.2 Types of connected elements

The shapes of the structural connected elements which are considered are:

- I or H columns (with a possible easy extension to RHS and CHS);
- I or H beams.

5.2.2.3 Types of loading

The design methods are intended for joints subjected to predominantly static or quasi-static loading. Fatigue aspects are not considered.

The resistance of the joints is checked under shear forces, correspondind to usual loading conditions of the structure during its life.

5.2.2.4 Types of fasteners

a)   Bolts

There are two classes of bolts: normal bolts and high strength bolts. The second class is characterised by a slip-type resistance mode in shear. Only non-preloaded bolts are covered below.

b)   Welds

In EN 1993-1-8, various types of welds are considered: fillet welds, fillet welds all round, butt welds, plug welds and flare groove welds. Only fillet welds are explicitly considered here.

5.2.2.5 Types of connections

Three connections types, used in the present design recommendations to connect a beam to a column or a beam to a beam, are specified below.

a)   Header plate connections

The main elements of a header plate connection are shown in Figure 5.6: a steel plate, a fillet weld on both sides of the supported beam web, and two single or two double vertical bolt lines. The plate is welded to the supported member and bolted to a supporting element such as a steel beam or column. Its height does not exceed the clear depth of the supported beam web. The end of the supported steel beam may be un-notched, single notched or double notched.

Figure 5.6 – Header plate connection

b)   Fin plate connections

The main elements of a fin plate connection are shown in Figure 5.7: a fin plate, a fillet weld on both sides of the plate, and a single or double vertical bolt line. The plate is welded to a supporting member such as a steel beam or column and bolted to the web of the supported beam. The end of the supported steel beam may be un-notched, single notched or double notched.

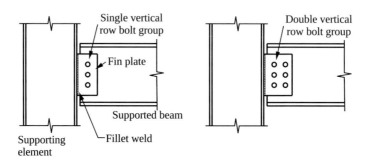

Figure 5.7 – Fin plate connection

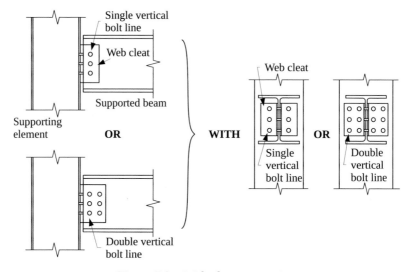

Figure 5.8 – Web cleat connection

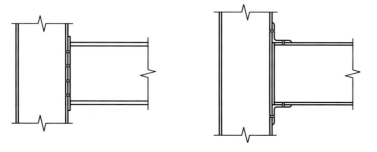

Figure 5.9 – Other simple connections

c)   Web cleat connections

A web cleat connection is characterised by two web cleats and three single or double vertical bolt lines (two on the supporting element and one on the supported member), see Figure 5.8. The cleats are bolted to the supporting and supported members. Un-notched, single notched or double notched supported beams may be considered.

Traditionally, other types of beam-to-column connections are sometimes considered as hinges. Current EN 1993-1-8 classifies them as semi-rigid. Two examples are given in Figure 5.9.

5.2.2.6 Reference code

The design rules presented in this chapter are based on the resistance formulae provided by EN 1993-1-8, at least as far as information is available. When this is not the case, the basic design principles given in EN 1993-1-8 are followed.

UK Specific Comment

"Header plates" are known as "partial depth end plates" in the UK. Web cleat connections are no longer commonly used in the UK. The full depth end plate connection shown on the left of Figure 5.9 is classified as a simple joint in the UK (SCI, 2011).

153

5.2.3 Joint modelling for frame analysis and design requirements

5.2.3.1 General

As explained in sub-chapter 2.2, the effects of the actual response of the joints on the distribution of internal forces and moments within a structure, and on the overall deformations, should generally be taken into account; but when these effects are sufficiently small, they may be neglected.

To identify whether the effects of joint behaviour on the analysis need to be taken into account, a distinction should be made between the three previously defined types of joint modelling: simple, continuous and semi-continuous. The appropriate type of joint modelling depends on the classification of the joint (see sub-chapter 2.4) and on the selected procedure for structural analysis and design.

5.2.3.2 Simple joint modelling

The joints considered in this chapter are assumed not to transmit bending moments. Thus, they should be modelled by hinges. Unfortunately, many joints which are traditionally considered as a hinge do not fulfil the stiffness and/or strength limitations required by EN 1993-1-8 for nominally pinned joints.

Two different approaches may be taken in such a case:

– According to the EN 1993-1-8 requirements, the joint is modelled as a rotational spring and is therefore considered as semi-rigid (what it is in reality). The rotational stiffness, bending resistance and shear resistance have to be evaluated and the actual properties of the joint have to be explicitly taken into consideration in the structural analysis and in the design phase. This approach is the most scientifically correct, but it needs more complex calculations as far as the global analysis and joint design are concerned;

– Despite its actual properties, the joint is considered as a hinge and the design rules presented in this chapter for simple joints can be applied, but under some strict conditions that ensure the safe character of the approach. The global analysis and the joint design are simpler in this case as they are based on a more traditional hinged (simple) approach.

If the second option is chosen, the joint is assumed not to transfer bending moments even if it is not absolutely true. Bending moments may develop in the joints although they are primarily designed to carry vertical shear forces. This is potentially unsafe and, at first sight, is not basically acceptable. But, a careful examination of this problem leads to the conclusion that the "hinge assumption" can be safe if the joint (i) has sufficient rotation capacity, and (ii) behaves ductile. The first requirement relates to the rotational capacity that the joint should have, in order to rotate as a hinge, without developing too high internal bending moments. The second requirement ensures that the development of combined shear and bending forces into the joint is not leading to brittle failure modes, such as rupture of a bolt or a weld. In other words, the design of the joint should exhibit internal plastic deformations instead of brittle phenomena.

If both requirements (sufficient rotation capacity and ductility) are fulfilled, it can be demonstrated that modelling a semi-rigid joint as nominally pinned is safe for design purposes and, in particular, for the evaluation of:

- *The frame displacements:* the stiffness of the actual structure is always greater than that of the hinged one, and all the actual displacements are therefore lower than the calculated ones;
- *The plastic failure load:* as the actual bending strength of the joint is higher than the assumed strength, taken as zero, the first order plastic resistance of the frame is higher that evaluated on the basis of a hinge behaviour;
- *The critical load of linear elastic instability:* the transversal stiffness of the actual structure is larger than the one of the structure with nominally pinned joints, and the rotational restraints at the end of the columns in the actual structure are higher than these calculated with a hinge assumption; this ensures the safe character of the hinge assumption, as far as global and local elastic instability are concerned;
- *The elastic-plastic phenomena of instability:* the actual stiffness of the structure is greater than the assumed stiffness but the actual loading conditions are more important than those acting on the structure with nominally pinned joints; nevertheless, various studies (Owens and Cheal, 1989; Gibbons *et al*, 1993; Gaboriau, 1995) show that the "hinged" approach is safe.

155

For further explanations, see Renkin (2003).

In this chapter, the design recommendations for simple joints relate to the "hinge model". Specific design requirements ensuring safety are presented for each of the connection types considered.

5.2.3.3 Summary of design requirements

As noted above, the internal forces in the joint are here determined by a structural analysis based on simple joint modelling. The hinge is assumed to be located at the intersection of the axes of the connected elements. As a result of this structural analysis, the maximum applied shear force and rotation in the joint, respectively $V_{j,Ed}$ and $\phi_{required}$ are obtained.

From the geometrical properties of the joint and the mechanical properties of its constitutive materials, the available rotation capacity of the joint $\phi_{available}$ can be estimated, as well as its design shear resistance $V_{j,Rd}$. To ensure the validity of this approach, some ductility requirements have to be satisfied and the available rotation of the joint has to be higher than the required. Finally, the joint will be considered as acceptable if the applied shear force does not exceed the design shear resistance.

UK Specific Comment

The UK Building Regulations require that joints are designed for tying forces in order to satisfy structural integrity requirements. Guidance on the design values of tying forces is given in BS EN 1991-1-7 Annex A and its UK NA. The requirements relate to the building class, with a design value of horizontal force generally not less than 75 kN, and usually significantly higher. Simple joints detailed to SCI (2011) satisfy these requirements.

5.2.4 Design resistance

The design of simple joints reduces to the limitation of the applied shear force to the shear design resistance, provided that ductility and rotation requirements are satisfied, see section 5.2.5. The evaluation of the shear design resistance is provided below for header plate and fin plate connections. The forces applied to joints at the ultimate limit state result from a structural analysis and has to be determined according to the principles given in EN 1993-1-1. The resistance of the joint is determined on the basis of the resistances of the individual fasteners, welds and other components, as shown below.

5.2.4.1 Connections with a header plate

5.2.4.1.1 Notations

Specific notations used in the following sections for header plates are shown in Figure 5.10.

Figure 5.10 – Header plate notations

5.2.4.1.2 Requirements to ensure the safety of the approach

To apply the design rules presented in section 5.2.4.1.3, all the following inequalities have to be satisfied.

1. $h_p \leq d_b$

2. $\dfrac{t_p}{h_e} > \phi_{required}$

3. If the supporting element is a beam or column web:

$$\frac{d}{t_p} \geq 2.8 \sqrt{\frac{f_{y,p}}{f_{ub}}} \quad \text{or} \quad \frac{d}{t_w} \geq 2.8 \sqrt{\frac{f_{y,w}}{f_{ub}}}$$

If the supporting element is a column flange:

$$\frac{d}{t_p} \geq 2.8 \sqrt{\frac{f_{y,p}}{f_{ub}}} \quad \text{or} \quad \frac{d}{t_{fc}} \geq 2.8 \sqrt{\frac{f_{y,fc}}{f_{ub}}}$$

4. $a \geq \dfrac{\beta_w \, f_{y,wb} \, \gamma_{M2}}{\sqrt{2} \, f_{u,wb} \, \gamma_{M0}} t_{wb}$

(β_w is given in Table 4.2)

5.2.4.1.3 Resistance to shear forces

The shear resistance of a header plate connection may be determined by considering the following eight components:

a) Bolts in shear

$$V_{Rd1} = 0.8\, n\, F_{v,Rd}$$

        n is the number of bolts;

        $F_{v,Rd}$ is given in Eq. (3.7) or (3.8).

b) Header plate in bearing

$$V_{Rd2} = n\, F_{b,Rd}$$

        $F_{b,Rd}$ is given in Eq. (3.16), with $f_u = f_{u,p}$, $t = t_p$, Figure 5.10.

c) Supporting member in bearing

$$V_{Rd3} = n\, F_{b,Rd}$$

        $F_{b,Rd}$ is, again, given in Eq. (3.16), with the relevant mechanical and geometrical properties.

Formulae as written here apply to major axis beam-to-column joints (connection to a column flange), to single-sided minor axis joints and to single-sided beam-to-beam joint configurations. In the other cases, the bearing forces result from both the left and right connected members, with the added problem that the number of connecting bolts may differ for the left and right connections. The calculation procedure may cover such cases without any particular difficulty. It could just bring some more complexity in the final presentation of the design sheet.

d) Header plate in shear: Gross section

$$V_{Rd\,4} = \frac{2h_p t_p}{1.27} \frac{f_{y,p}}{\sqrt{3}\,\gamma_{M0}} \qquad \text{(2 resisting sections)}$$

e) Header plate in shear: Net section

$$V_{Rd\,5} = 2A_{v,net} \frac{f_{u,p}}{\sqrt{3}\,\gamma_{M2}} \qquad \text{(2 resisting sections)}$$

    with $A_{v,net} = t_p \left(h_p - n_1 d_0 \right)$ and n_1: number of horizontal bolt rows.

f) Header plate in shear: Shear block

$$V_{Rd6} = 2\, F_{eff,Rd}$$

                (2 resisting sections)

- If $h_p < 1.36\, p_{22}$ and $n_1 > 1$:

$$F_{eff,Rd} = F_{eff,2,Rd} = 0.5 \frac{f_{u,p} A_{nt}}{\gamma_{M2}} + \frac{1}{\sqrt{3}} f_{y,p} \frac{A_{nv}}{\gamma_{M0}}$$

- else:

$$F_{eff,Rd} = F_{eff,1,Rd} = \frac{f_{u,p} A_{nt}}{\gamma_{M2}} + \frac{1}{\sqrt{3}} f_{y,p} \frac{A_{nv}}{\gamma_{M0}}$$

with $\qquad p_{22} = p_2'$ $\qquad\qquad$ for $n_2 = 2$

$\qquad\qquad p_{22} = p_2' + p_2$ $\qquad\qquad$ for $n_2 = 4$

n_2 is the number of vertical bolt rows

A_{nt} = net area subjected to tension

- for one vertical bolt row ($n_2 = 2$):

$$A_{nt} = t_p \left(e_2 - d_0/2 \right)$$

- for two vertical bolt rows ($n_2 = 4$):

$$A_{nt} = t_p \left(p_2 + e_2 - 3d_0/2 \right)$$

A_{nv} = net area subjected to shear

$$A_{nv} = t_p \left[h_p - e_1 - \left(n_1 - 0.5 \right) d_0 \right]$$

(see clause 3.10.2 in EN 1993-1-8)

159

g) Header plate in bending

- If $h_p \geq 1.36\, p_{22}$: $\qquad V_{Rd7} = \infty$

- else: $\qquad\qquad V_{Rd7} = \dfrac{2 W_{el}}{\dfrac{\left(p_{22} - t_w \right)}{2}} \dfrac{f_{y,p}}{\gamma_{M0}}$

with $W_{el} = \dfrac{t_p h_p^2}{6}$

h) Beam web in shear

$$V_{Rd8} = t_{wb}\, h_p\, \frac{f_{y,wb}}{\gamma_{M0} \sqrt{3}} \qquad \text{(clause 5.4.6 in EN 1993-1-1)}$$

i)   Shear resistance of the joint

$$V_{Rd} = \min_{i=1}^{8}\{V_{Rd\,i}\}$$

5.2.4.2 Connections with a fin plate

5.2.4.2.1 Notations

Specific notations for fin plate connections used in the following sections are shown in Figure 5.11.

Figure 5.11 – Fin plate notations

5.2.4.2.2 Requirements to ensure sufficient rotation capacity

The two following inequalities have to be satisfied.

1. $h_p \le d_b$

2. $\phi_{available} > \phi_{required}$, see section 5.2.5.2.1.

5.2.4.2.3 Requirements to avoid premature weld failure

The following inequality has to be fulfilled.

3. $a \ge \dfrac{\beta_w \; f_{y,p} \; \gamma_{M2}}{\sqrt{2}\; f_{u,p} \; \gamma_{M0}} t_p$

(β_w is given in Table 4.2)

5.2.4.2.4 Resistance to shear forces

a)   Bolts in shear

$$V_{Rd1} = \frac{n\,F_{v,Rd}}{\sqrt{1+\left(\dfrac{6\,z}{(n+1)\,p_1}\right)^2}} \qquad\qquad \text{for } n_2 = 1$$

$$V_{Rd1} = \frac{F_{v,Rd}}{\sqrt{\left(\dfrac{z\,p_2}{2\,I}+\dfrac{1}{n}\right)^2+\left(\dfrac{z\,p_1}{2\,I}\,(n_1-1)\right)^2}} \qquad \text{for } n_2 = 2$$

with:

$$I = \frac{n_1}{2}\,p_2^{\,2}+\frac{1}{6}\,n_1\left(n_1^{\,2}-1\right)p_1^{\,2}$$

and $F_{v,Rd}$ is given in Eq. (3.7) or (3.8).

b)   Fin plate in bearing

$$V_{Rd2} = \frac{1}{\sqrt{\left(\dfrac{\dfrac{1}{n}+\alpha}{F_{b,ver,Rd}}\right)^2+\left(\dfrac{\beta}{F_{b,hor,Rd}}\right)^2}}$$

-    for $n_2 = 1$:

$$\alpha = 0 \text{ and } \beta = \frac{6\,z}{p_1\,n(n+1)}$$

-    for $n_2 = 2$:

$$\alpha = \frac{z}{I}\frac{p_2}{2} \text{ and } \beta = \frac{z}{I}\frac{n_1-1}{2}\,p_1$$

$F_{b,ver,Rd}$ and $F_{b,hor,Rd}$ are given in Eq. (3.16), with the relevant mechanical and geometrical properties.

c)   Fin plate in shear: Gross section

$$V_{Rd3} = \frac{h_p t_p}{1.27} \frac{f_{y,p}}{\sqrt{3}\, \gamma_{M0}}$$

d)   Fin plate in shear: Net section

$$V_{Rd4} = A_{v,net} \frac{f_{u,p}}{\sqrt{3}\, \gamma_{M2}}$$

and $A_{v,net}$ is defined in 5.2.4.1.3 e).

e)   Fin plate in shear: Shear block

$$V_{Rd5} = F_{eff,\,2,Rd}$$

and $F_{eff,2,Rd}$ is defined in 5.2.4.1.3 f).

f)   Fin plate in bending

–   If $h_p \geq 2.73\ z$: $V_{Rd6} = \infty$

–   else: $V_{Rd6} = \frac{W_{el}}{z} \frac{f_{yp}}{\gamma_{M0}}$

with $W_{el} = \dfrac{t_p h_p^2}{6}$

g)   Buckling of the fin plate

Derived from BS 5950-1 (BSI, 2000; BCSA/SCI Connections Group *et al*, 2002):

$$V_{Rd7} = \frac{W_{el}}{z_p} \frac{f_{pLT}}{0.6\gamma_{M1}} \leq \frac{W_{el}}{z_p} \frac{f_{yp}}{\gamma_{M0}} \qquad \text{if } z_p > \frac{t_p}{0.15}$$

$$V_{Rd7} = V_{Rd6} \qquad \text{if } z_p \leq \frac{t_p}{0.15}$$

where:

f_{pLT} is   the lateral torsional buckling strength of the plate, see Annex B

h)   Beam web in bearing, computed as in 5.2.4.2.4 b) using the relevant mechanical and geometrical properties.

i)   Beam web in shear: Gross section, computed as in 5.2.4.1.3 h).

j)   Beam web in shear: Net section

$$V_{Rd\,10} = A_{b,v,net} \frac{f_{u,wb}}{\sqrt{3}\,\gamma_{M2}}$$

with $A_{b,v,net} = A_{b,v} - n_1 d_0 t_{bw}$

k)   Beam web in shear: Shear block, computed as in 5.2.4.2.4 e) using the relevant mechanical and geometrical properties

l)   Shear resistance of the joint

$$V_{Rd} = \min_{i=1}^{11}\{V_{Rd\,i}\}$$

5.2.5 Practical ways to satisfy the ductility and rotation requirements

The interested reader will find detailed information on the ways to satisfy the ductility and rotation requirements in Jaspart *et al* (2009). In the next sections, these requirements are illustrated for every connection type.

5.2.5.1 Header plate connection

5.2.5.1.1 Design requirements to ensure sufficient rotation capacity

To enable rotation without increasing too much the bending moment which develops into the joint, contact between the lower beam flange and the supporting member has to be strictly avoided. So, it is imperative that the depth h_p of the plate is less than that of the supported beam web (Figure 5.12):

$$h_p \le d_b \tag{5.1}$$

where d_b is the clear depth of the supported beam web.

If contact occurs, a compression force develops at the place of contact; it is equilibrated by tension forces in the bolts and a significant bending moment develops (Figure 5.12).

163

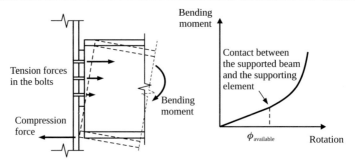

Figure 5.12 – Contact and evolution of the bending moment

The level of rotation at which contact occurs is obviously dependent on the geometrical characteristics of the beam and header plate, but also on the actual deformations of the joint components.

In order to derive a simple criterion that the user could apply, before any calculation, to check whether the risk of contact may be disregarded, the following rough assumptions are made (see Figure 5.13):

- The supporting element remains un-deformed;
- The centre of rotation of the beam is located at the lower extremity of the header plate.

Based on these assumptions, a safe estimation, i.e. a lower bound, of the available rotation of the joint may be easily derived:

$$\phi_{available} = \frac{t_p}{h_e} \qquad (5.2)$$

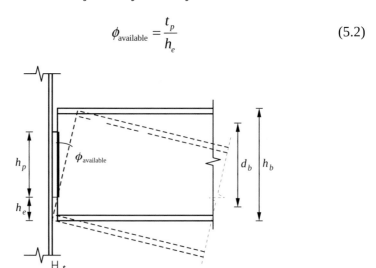

Figure 5.13 – Geometrical characteristics of the joint and illustration of contact between the beam and the supporting element

The available rotation has to be greater than the required rotation capacity, which varies according to the structural system and loading:

$$\phi_{available} > \phi_{required} \qquad (5.3)$$

For instance, the required rotation capacity, for a beam (length L and second moment of area I) simply supported at its extremities and subjected to an uniformly distributed load (factored load γp at ULS), is given by:

$$\phi_{required} = \frac{\gamma p L^3}{24EI} \qquad (5.4)$$

By expressing that $\phi_{required} > \phi_{available}$, a simple criterion ensuring a sufficient joint rotation capacity may be derived:

$$\frac{t}{h_e} > \frac{\gamma p L^3}{24EI} \qquad (5.5)$$

Similar criteria may be derived for other load cases (Annex A).

5.2.5.1.2 Design requirements to ensure sufficient joint ductility

As bending moments develop in the joint, the bolts and the welds are subjected to tension forces in addition to shear forces. Premature failure of those elements, which exhibit a brittle failure, and which are more heavily loaded in reality than in the calculation model, has therefore to be strictly avoided. Simple related criteria should therefore be proposed.

165

a)   Criterion to avoid premature bolt failure because of tension forces

In Eurocode 3, a criterion based on the T-stub approach ensures that a yield line mechanism develops in the plate before the strength of the bolts is exhausted, see EN 1993-1-8; its background is given in Jaspart (1997). This criterion, initially developed for end plates and column flanges, is here safely extended to column (weak axis beam-to-column joints) or beam (beam-to-beam joint configurations) webs.

According to this criterion, at least one of the two following Eqs. (5.6) to (5.8) has to be satisfied:

– For header plate:

$$\frac{d}{t_p} \geq 2.8\sqrt{\frac{f_{y,p}}{f_{ub}}} \tag{5.6}$$

– For a supporting column flange:

$$\frac{d}{t_{cf}} \geq 2.8\sqrt{\frac{f_{y,cf}}{f_{ub}}} \tag{5.7}$$

– For a supporting column or beam web (or faces of hollow sections):

$$\frac{d}{t_w} \geq 2.8\sqrt{\frac{f_{yw}}{f_{ub}}} \tag{5.8}$$

where:

d	is	the nominal diameter of the bolt;
t_p	is	the thickness of the header plate;
t_{cf}	is	the thickness of the supporting column flange;
t_w	is	the thickness of the supporting column or beam web;
$f_{y,p}$	is	the yield strength of the steel constituting the header plate;
$f_{y,cf}$	is	the yield strength of the steel constituting the supporting column flange;
f_{yw}	is	the yield strength of the steel constituting the supporting column or beam web;
f_{ub}	is	the ultimate strength of the bolt.

Such a criterion does not ensure that the whole shear capacity of the bolt may be considered when evaluating the shear resistance of the joint. In fact, when this requirement is satisfied, it may be demonstrated that:

– the tension force in the bolts may be up to $0.5F_{t,Rd}$, i.e. 50% of the design tension resistance $F_{t,Rd}$ of the bolts, but no more;
– for such a maximum tension force, the actual shear resistance of the bolt is only 64% of the full shear resistance of the bolts (according to the EN 1993-1-8 resistance formula for bolts in shear and tension).

This looks disappointing, at first sight: the user tries to maximise the shear resistance of the joint but should limit the resistance of the bolts in shear by 36%. It may be argued though that only the bolts located in the upper half of the header plane are affected by such a reduction, as the others are located in a compression zone, and are therefore not subjected to tension forces.

So finally a reduction is taken into consideration by multiplying the total resistance of the bolts in shear by a factor of 0.8, i.e. a reduction factor of 0.64 for half of the bolts located in the upper half of the header plate, $0.5×(1+0.64) ≈ 0.8$.

b)   Criterion to avoid premature weld failure because of tension or shear forces

The welds must be designed according to EN 1993-1-8. In the case of relatively small loads in relation to the capacity of the web, application of the rules in 4.5.3.2 of EN 1993-1-8 may lead to rather thin welds. If the ultimate strength of those thin welds is lower than the yield strength of the weakest of the connected parts, the connection has so little deformation capacity that is usually not sufficient to accommodate imposed deformations, etc. In such a case the connection will behave in a brittle way.

To avoid this, the welds should be designed full strength. The ultimate strength of full strength welds would then be greater than the ultimate strength of the adjacent plate; so, in the case of overloading, the plate would yield before the welds fail.

In section 4.3.6, the concept of "full strength welds" has been introduced and related design criteria have been derived. These criteria may be expressed separately for welds in shear and welds in tension. But in the present case, the welds are subjected simultaneously to shear and tension forces; as a result, the worst situation in terms of weld loading is therefore considered for design: end welds, i.e. welds subjected to pure tension.

By applying clause 4.5.3.2 of EN 1993-1-8, using the directional method, the minimum throat thickness value required for a full strength weld design are obtained, see Table 4.3.

5.2.5.2 Fin plate connection

5.2.5.2.1 Design requirements to ensure sufficient rotation capacity

To avoid significant moment transfer as a joint rotates, contact between the lower beam flange and the supporting member has to be strictly avoided. To achieve this, the height h_p of the fin plate should be lower than that of the supported beam web (Figure 5.14):

$$h_p \leq d_b \tag{5.9}$$

If contact takes place, a compression force develops at the place of contact; it is equilibrated by tension forces in the welds and in the plate, and additional shear forces in the bolts.

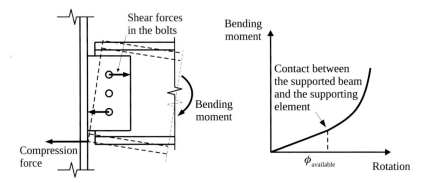

Figure 5.14 – Contact and evolution of the bending moment

The level of rotation at which contact occurs is obviously dependent on the geometrical characteristics of the beam and of the fin plate, but also on the actual deformations of the joint components.

In order to derive a simple criterion that the user could apply, before any calculation, to check whether the risk of contact may be disregarded, the following rough assumptions are made (see Figure 5.15):

   – The supporting element and the fin plate remain undeformed;
   – The centre of rotation of the beam is located at the centre of gravity of the bolt group.

A safe estimation, i.e. a lower bound, of the available rotation of the joint may then be easily derived:

$$\text{if } z > \sqrt{\left(z - g_h\right)^2 + \left(\frac{h_p}{2} + h_e\right)^2} :$$

$$\phi_{available} = \infty \tag{5.10}$$

else:

$$\phi_{available} = \arcsin\left(\frac{z}{\sqrt{(z-g_h)^2+\left(\frac{h_p}{2}+h_e\right)^2}}\right) - \text{atan}\left(\frac{z-g_h}{\frac{h_p}{2}+h_e}\right) \qquad (5.11)$$

Expressions for $\phi_{required}$ are given in section 5.2.5.1.1 and Annex A.

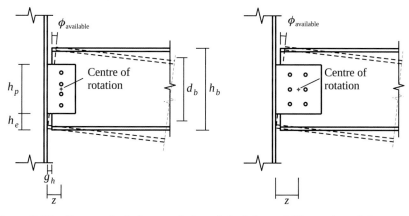

Figure 5.15 – Geometrical characteristics of the joint and illustration of the contact between the beam and the supporting element

5.2.5.2.2 Design requirements to ensure sufficient joint ductility

169

As previously explained, the design shear resistance of the joint may be reached, as a result of a plastic redistribution of internal forces among the different constitutive components. This requires that no local brittle failure modes or instabilities develop during this redistribution. For fin plate connections, the failure modes which could prevent redistribution of internal forces to take place are: the bolts and the welds in shear because of their brittle nature, and the buckling of the fin plate which is assumed to be non-ductile in terms of plastic redistribution.

a)   Criterion to avoid premature weld failure because of tension forces

The same full strength criterion as the one established for the header plate connection may be considered (see Section 5.2.5.1.2). Related minimum weld throat thicknesses can be obtained from Table 4.3.

b) Criterion to permit a plastic redistribution of internal forces between the "actual" and "design" loading situations

First of all, the design shear resistance of the connection should be associated with a ductile mode. Failure by bolts in shear or by buckling of the fin plate is therefore excluded. A first criterion can be written as:

$$\min\left(V_{Rd\,1}; V_{Rd\,7}\right) > V_{j,Rd} \tag{5.12}$$

where:

$V_{Rd\,1}$ is the shear resistance of the bolts, see section 5.2.4.2.4a);

$V_{Rd\,7}$ is the buckling resistance of the fin plate, see section 5.2.4.2.4g).

Secondly, the component that yields under the "actual" loading in the connection (which will not normally be shear alone) has also to be ductile, so, failure governed by the bolts in shear or buckling of a fin plate has to be precluded. To ensure this, different criteria have to be fulfilled dependent on the failure mode obtained through treating the connections as hinged:

– Failures by bolts in shear or buckling of the fin plate: excluded by the first criterion, Eq. (5.12);

– All other failure modes: for one vertical bolt row, at least one of the following two inequalities has to be satisfied:

For the beam web:

$$F_{b,hor,Rd} \leq \min\left(V_{Rd\,1}; V_{Rd\,8}\beta\right) \tag{5.13}$$

For the fin plate:

$$F_{b,hor,Rd} \leq \min\left(V_{Rd\,1}; V_{Rd\,7}\beta\right) \tag{5.14}$$

β is defined in section 5.2.4.2.4. For two vertical bolt rows, at least one of the following three inequalities has to be satisfied:

1. For the beam web:

$$\max\left(\frac{1}{F_{v,Rd}^2}\left(\alpha^2 + \beta^2\right); \frac{1}{V_{Rd\,7}^2}\right) \leq \left(\frac{\alpha}{F_{b,ver,Rd}}\right)^2 + \left(\frac{\beta}{F_{b,hor,Rd}}\right)^2 \tag{5.15}$$

2. For the fin plate:

$$\max\left(\frac{1}{F_{v,Rd}^2}(\alpha^2+\beta^2);\frac{1}{V_{Rd\,7}^2}\right)\leq\left(\frac{\alpha}{F_{b,ver,Rd}}\right)^2+\left(\frac{\beta}{F_{b,hor,Rd}}\right)^2 \qquad (5.16)$$

3. And:

$$V_{Rd\,6}\leq\min\left(\frac{2}{3\sqrt{\alpha^2+\beta^2}}F_{v,Rd};\frac{2}{3}V_{Rd\,7}\right) \qquad (5.17)$$

Finally, during the redistribution process, the bolts in shear failure mode should be prevented. Simple criteria can be written that again depends on the failure mode resulting from treating the connection as a "hinge":

– Failure by bolts in shear or buckling of the fin plate: excluded by the first criterion (1).
– Failure by fin plate or beam web in bearing: If the two first criteria (1) and (2) are fulfilled, no additional criterion is necessary.
– All other failure modes:

$$V_{Rd\,1}>\min\left(V_{Rd\,2};V_{Rd\,8}\right) \qquad (5.18)$$

where: V_{Rd2} is the bearing resistance of the fin plate;
           V_{Rd8} is the bearing resistance of the beam web.

The criteria (1), (2) and (3) can be only checked after the evaluation of the design shear resistance of the joint. For further explanations about the derivation of these requirements, see Renkin (2003).

5.2.5.3 Web cleat connection

In Jaspart *et al* (2009), web cleat connections are just briefly addressed as it is stated that "the components to be considered are simply those found in header plate (cleat leg connected to the supporting member) and fin plate (cleat leg connected to the supported member) connections". It is so concluded that the design rules and requirements for a safe approach may be simply deduced from those established for the two previous connection types. In Weynand *et al* (2012) however it is pointed out that the previous statement is true for all components active in a web cleat connection except for the consideration of *header plate in bending*, which requires further consideration. The conclusions drawn in Weynand *et al*, (2012) are summarised below.

In Jaspart *et al* (2009), the header plate is modelled as a simple single-span beam with elastic stress distribution over its rectangular cross section. This model is not applicable to web cleats. In a double web cleated connection, the eccentricity of the bolts connected to the supporting member lead to a moment M_y, see Figure 5.16, resulting in compression forces in the upper part and horizontal tension forces in the lower part of the cleat. It is assumed that the compression forces will be carried by contact of the cleats with the web of the beam (Figure 5.17). The width of the compression zone b_{com} is derived by assuming an angle $\alpha = 60°$. The height of the compression zone h_{com} results from the criterion that the compression force should not exceed the yield strength of the plates (Figure 5.18).

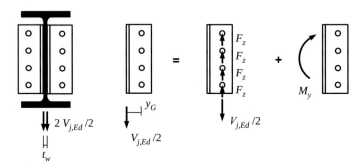

Figure 5.16 – Forces at supporting member side

Figure 5.17 – Compression zone

Figure 5.18 – Forces on the supporting element side

The tension forces are taken by additional (horizontal) forces in the bolts of the leg connected to the supporting member (Figure 5.18), leading to a reduction in the vertical component of the shear force in the bolt.

The centre of compression is assumed to be located at mid height of the compression zone. An elastic distribution of the resulting horizontal forces is assumed. Hence, horizontal forces in the bolts increase linearly (Figure 5.18). Under these assumptions and starting with an initial guess for the height of the compression zone h_{com}, all forces may be determined and a more accurate value of h_{com} can be found by a short iteration.

Because of the short distance from the compression zone to the supporting element side, the horizontal forces are directly resisted by the bolts at the supporting element side (Figure 5.17). Torsion in the cleat and additional tying forces in the bolts on the supported element side do not arise.

In order to avoid notched beams, a configuration with double web cleated connections with "long" legs may be used as shown in Figure 5.19. To check the "bending in the cleat" connected to the supporting member, a compression zone as shown in Figure 5.17 may not be assumed. Hence, the cleat can rotate more freely and the moment M_y, see Figure 5.16, must be carried by the bolt pattern of the cleat legs on the supporting member. Resulting forces are shown in Figure 5.19. If these bolts in shear are the decisive component of the joint, the presence of the horizontal forces due to M_y will obviously lead to a reduction of the shear resistance of the joint.

173

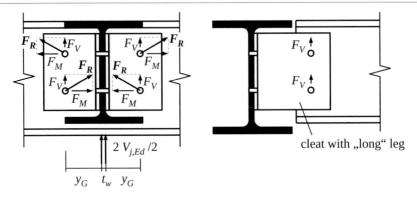

Figure 5.19 – Forces on the supporting element side for cleats with long legs

---

UK Specific Comment

It is generally safe to assume that the ductility (rotation requirements) are satisfied provided that joints are detailed in accordance with SCI (2011).

5.3   COLUMN BASES

5.3.1 Introduction

Simple column bases are assumed to transfer axial (compression or tension) and shear forces. Tension forces may result from an uplift of the structure.

In Figure 5.20 two classical types of column bases, respectively with two and four anchor bolts, are illustrated. The first type, with two anchor bolts, meets more closely the concept of simple joint than the second one which may develop significant moments, as a result of the lever arm provided by the two lines of anchor bolts. Obviously, the use of rather thin and unstiffened base plates in the "four anchors" configuration is likely to limit the transfer of bending to the concrete block.

But again here, as stated in section 5.2.3.2, this actual semi-rigidity of the column bases will never prevent the designer from considering them as simple as long as the joint exhibits sufficient rotational capacity and ductility. To achieve these, measures will have to be taken so that anchor bolt failure and weld failure are not governing failure modes, see section 5.2.5.1.2.

Figure 5.20 – Classical column base detailing, configuration with two and four anchor bolts

Typical anchor bolt types are illustrated in Figure 5.21, namely hooked bars for light anchoring, cast-in-place headed anchors and anchors bonded to drilled holes.

a) cast in place   c) adhesive   e) expansion

b) post-installed   d) grouted   f) anchorage to grillage systems

Figure 5.21 – Various types of anchoring systems

When it is necessary to transfer a large force, more expensive anchoring systems such as grillage beams embedded in concrete are to be selected.

In the next sections, the application of the component method to simple column bases according to EN 1993-1-8 is presented. The interested reader may refer to Wald *et al* (2008a, 2008b), Steenhuis *et al* (2008), Jaspart *et al* (2008b) and Gresnigt *et al* (2008) for more detailed information.

175

> **UK Specific Comment**
>
> The column base detailing on the left of Figure 5.21 is not commonly used in the UK. We never recommend less than four bolts – they are needed to assure stability and facilitate location during erection. The anchoring systems shown in Figure 5.22 are not usually adopted in the UK. Reference is made to SCI (2013a) for common anchor systems.

5.3.2 Basis for the evaluation of the design resistance

According to the component method, the evaluation of the design resistance of joints, in this case under axial compression/tension and shear forces, requires the identification, the characterisation and the assembly of the constitutive joint components.

The active components in simple column base joints are illustrated in Figure 5.22. The components base plate and concrete block in compression, and the base plate in bending and anchor bolts in tension contribute to the transfer of axial force through the joint, respectively under compression and tension. The anchors bolt in shear relates to the transfer of shear force.

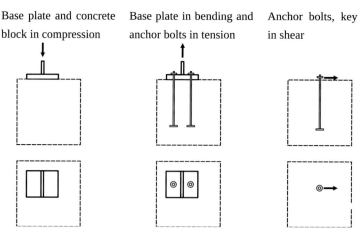

Figure 5.22 – Components in a simple column base

In the following pages, the aspects of characterisation and assembly are addressed and the resulting design procedures for simple column bases under axial and shear forces are respectively presented.

5.3.3 Resistance to axial forces

5.3.3.1 Component "base plate and concrete block in compression"

When considering the base plate and concrete block in compression component, the grout layer located between the base plate and the concrete is considered, as it influences the resistance of the component.

To model the response of the base plate in compression, two approaches are found in the literature:

- – The base plate is assumed to be rigid in bending;
- – The base plate is assumed to be flexible in bending.

The difference between rigid and flexible plates can be explained using a base plate connection loaded by an axial force only. In the case of rigid plates it is assumed that the stresses under the plate are uniformly distributed. In the case of flexible plates, the stresses are concentrated around the footprint of the column section under the plate. In EN 1993-1-8, the flexible model is considered as (i) it better corresponds to the reality, and as (ii) it has been explained before that the need for rotational ductility in simple column bases implies the use of unstiffened and not too thick plates (especially for configurations with four anchor bolts).

Various researchers (Shelson, 1957; Hawkins, 1968b, 1968a; DeWolf, 1978) experimentally investigated the resistance of the base plate and concrete block in compression component. Factors influencing this resistance are the concrete strength, the plate area, the plate thickness, the grout, the location of the plate on the concrete foundation, the size of the concrete foundation and reinforcement. Concerning modelling, Stockwell (1975) introduced the concept of replacing a flexible plate with a non-uniform stress distribution by an equivalent rigid plate with a uniform stress distribution. Steenhuis and Bijlaard (1999), and Murray (1983) verified this simple practical method with experiments and suggested improvements. EN 1993-1-8 has adopted this method in a form suitable for standardisation. This method is described below.

The resistance is thus determined by means of an equivalent rigid plate concept. Figure 5.23 shows how an equivalent rigid plate can replace a flexible plate when the base plate connection is loaded by axial force only. In this figure, A is the area of top surface of the concrete block, A_p is

177

the area of the plate, A_{eq} is the area of the equivalent rigid plate, and c is used to determine the equivalent width (or length) of footprint. The resistance is determined by two parameters: the bearing strength of the concrete and the dimensions of the equivalent rigid plate.

a)   Bearing strength of the concrete

The bearing strength of the concrete underneath the plate depends on the size of the concrete block. The edge effect is taken into account by the following definition of the concentration factor

$$k_j = \sqrt{\frac{a_1\, b_1}{a\, b}} \tag{5.19}$$

where the geometrical edge conditions, see Figure 5.24, are introduced by

$$a_1 = \min\left\{ \begin{array}{l} a + 2\, a_r \\ 5\, a \\ a + h \\ 5\, b_1 \end{array} \right\}, \quad a_1 \geq a \tag{5.20}$$

$$b_1 = \min\left\{ \begin{array}{l} b + 2\, b_r \\ 5\, b \\ b + h \\ 5\, a_1 \end{array} \right\}, \quad b_1 \geq b \tag{5.21}$$

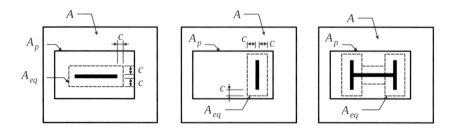

Figure 5.23 – Flexible base plate modelled as a rigid plate of equivalent area

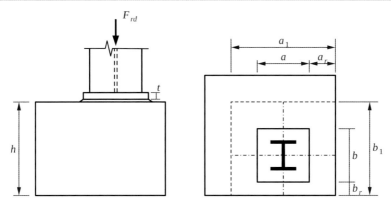

Figure 5.24 – Concrete block geometrical dimensions

If there is no edge effect, it means that the geometrical position of the column base is sufficiently far away from the edges of the concrete block and the value for k_j according to EN 1993-1-8 is 5. This concentration factor is used for the evaluation of the design value of the bearing strength as follows

$$f_{j,d} = \frac{\beta_j \, k_j \, f_{c,k}}{\gamma_c} \tag{5.22}$$

where γ_c is a partial safety factor for concrete. A reduction factor β_j is used to take into account that the resistance under the plate might be smaller due to the quality of the grout layer. The value $\beta_j = 2/3$ may be used if the grout characteristic strength is more than 0.2 times the characteristic strength of the concrete foundation, $f_{c,g} \geq 0.2 f_{c,k}$, and the thickness of the grout is smaller than 0.2 times the minimum base plate width, $t_g \leq 0.2 \min(a,b)$. In case where the thickness of the grout is more than 50 mm, the characteristic strength of the grout should be at least the same as that of the concrete foundation. These conditions are usually fulfilled. If not, the grout should be checked separately, see Steenhuis *et al* (1999).

b)   Dimensions of the equivalent rigid plate

The flexible base plate, with area A_p, can be replaced by an equivalent rigid plate with area A_{eq}. This rigid plate area A_{eq} is built up from one T-stub under the column web and two T-stubs under the column flanges. For the T-stub illustrated in Figure 5.25, its equivalent rigid area equals $a_{eq,R} \cdot L$.

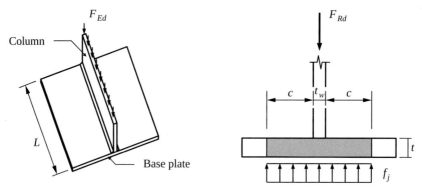

Figure 5.25 – T-stub under compression

The equivalent width c of the T-stub under compression, see Figure 5.25, can be determined assuming that no plastic deformations will occur in the flange of the T-stub. Therefore, the resistance per unit length of the T-stub flange is taken as the elastic resistance:

$$M' = \frac{1}{6}t^2 f_y \qquad (5.23)$$

It is assumed that the T-stub is loaded by a uniform stress distribution. The bending moment per unit length in the base plate acting as a cantilever with span c is:

$$M' = \frac{1}{2}f_{j,d}c^2 \qquad (5.24)$$

where $f_{j,d}$ is concrete bearing strength. The equivalent width c can be resolved by combining Eqs. (5.23) and (5.24):

$$c = t\sqrt{\frac{f_y}{3f_{j,d}\gamma_{M0}}} \qquad (5.25)$$

The width of the T-stub is now:

$$a_{eq,R} = t_w + 2c = t_w + 2t\sqrt{\frac{f_y}{3f_{j,d}\gamma_{M0}}} \qquad (5.26)$$

The resistance F_{Rd} of the T-stub, see Figure 5.25, should be higher than the applied load F_{Ed}:

$$F_{Ed} \le F_{Rd} = a_{eq,R} \cdot L \cdot f_{j,d} \qquad (5.27)$$

The base plate is stiffer in bending near the intersections of web and flanges. This stiffening effect is not taken into consideration when computing the equivalent area A_{eq}. Wald (1993) shows that this stiffening effect may result in a 3% higher resistance for open sections and a 10% higher resistance for tubular sections in comparison to the method of determination of A_{eq}. The calculation of the concentration factor k_j based on Eq. (5.19) leads to conservative results. This can be improved by modification of the procedure of Eqs. (5.19) – (5.21). In that case, the equivalent area instead of the full area of the plate should be considered. However, this iterative procedure is not recommended for practical purposes. In case of high quality grout, a less conservative procedure with a distribution of stresses under 45° may be adopted, see Figure 5.26.

Figure 5.26 – Stress distribution in the grout

The bearing stress under the plate increases with a larger eccentricity of the axial force (DeWolf and Ricker, 1990; Penserini and Colson, 1989). In this case, the base plate is in larger contact with the concrete block due to its bending and the stress in the edge under the plate increases. However, the effect of these phenomena is limited.

The influence of packing under the steel plate may be neglected for practical design (Wald *et al*, 1993). The influence of the washer under plate used for construction can be also neglected in case of grout quality

$f_{c,g} \geq 0.2f_c$. The anchor bolts and base plate resistance should be taken into account explicitly in case of grout quality $f_{c,g} < 0.2f_c$.

5.3.3.2 Component base plate in bending and anchor bolts in tension

Anchor bolts are longer compared to the bolts used in end plates due to presence of washer plates and grout, thick base plate and the part embedded in the concrete block. As a result, the elongation of the anchor bolts could lead to the separation of the base plate from the concrete block when the anchor bolts are loaded in tension. Different responses are therefore to be considered when the strength of the base plate loaded by bending moment and/or tension axial force has to be predicted.

The failure of the base plate in bending and anchor bolts in tension component may result from the yielding of the plate, the failure of the anchor bolts or a combination of both phenomena.

Figure 5.27 – T-stub idealisation (case with prying effects)

In order to model the base plate in bending and the bolts in tension, reference is again made to the so-called "equivalent T-stub". This is illustrated in Figure 5.27 for one of the extended parts of the base plate (and the corresponding anchor bolts).

If the bolts elongate significantly and the plate is stiffened, the plate separates from the concrete foundation. Otherwise, the edge of the plate is in contact with the concrete resulting, what results in the development of prying forces Q (see Figure 5.27).

In section 3.5.2, the design formulae for T-stubs in tension are provided, as well as the criterion which allows defining whether prying effects have to be or not considered. The boundary criterion is:

$$L_{b,lim} = \frac{8.8 \; m^3 A_s}{l_{eff} \, t^3} \tag{5.28}$$

where $L_{b,lim}$ and A_s designate respectively the effective length and the cross section design area of the anchor bolts. l_{eff} is the length of the equivalent T-stub (see section 3.5.2).

The use of the T-stub concept requires the preliminary evaluation of the design resistance of the anchors in tension and of the length of the anchors.

As far as resistance is concerned, the lack of anchoring (e.g. pull-out of the anchor, failure of the concrete) should be avoided and collapse of anchor bolts be preferred. This avoids brittle failure of the anchoring. For seismic areas, the failure of the column base should occur in the base plate rather than in the anchor bolts (Astaneh *et al*, 1992). The plastic mechanism in the plate ensures ductile behaviour and dissipation on energy.

For a single anchor, the following failure modes have to be considered:

- The pull-out failure $N_{p,Rd}$;
- The concrete cone failure $N_{c,Rd}$;
- The splitting failure of the concrete $N_{sp,Rd}$.

Similar verification is required for a group of anchor bolts.

The detailed description of the related design formulae is included in the CEB Guide (CEB, 1994) for various types of fastening.

In the case of embedded anchor bolts, the effective length of the bolt L_b consists of the free length L_{bf} and an effective embedded length L_{be}, $L_b = L_{bf} + L_{be}$, see Figure 5.28.

Figure 5.28 – Effective length of an embedded anchor bolt

According to EN 1993-1-8, the effective embedded length of bolts with regular surface is assumed to be equal to $L_{be} \approx 8d$. The minimum actual embedded length of the anchor allowing to justify the use $L_{be} \approx 8d$ is $24d$. This value has therefore to be considered as the minimum actual embedded length of the anchor bolts. This length may be reduced by the use of a headed plate (as illustrated in Figure 5.20).

5.3.3.3 Assembly of components for resistance evaluation

The simple joints under consideration in the present section are illustrated in Figure 5.20: base plates with two or four anchor bolts. Except in cases where the uplift of the structure would induce high tension forces in the column, only two anchors bolts will be used so as to limit the actual transfer of bending moments mentioned in section 5.3.1, which results from the fact that the actual joints do not have a "perfectly simple" behaviour.

Whatever the number of anchor bolts, the configuration of the joint is assumed to be symmetrical, with reference to the axis of the column. In fact, the joint is not symmetric, the application of an axial force in the column will result in the development of an axial force and a bending moment in the joint. These aspects of combined loading in the joint will be addressed later in chapter 6.

The axial resistance of a simple column base joint, respectively under axial compression and tension forces, may be computed as follows:

a)   Design resistance in compression

$$N_{Rd} = A_{eq} f_{j,d} \qquad (5.29)$$

b)   Design resistance in tension

$$N_{Rd} = \sum_{i}^{n} F_{T,Rd,i} \tag{5.30}$$

where:

$F_{T,Rd,i}$ is the design resistance of the individual T-stub i, see section 5.3.2; the summation extends to all the n individual T-stubs which may be identified in the base plate (one per bolt-row).

5.3.4 Resistance to shear forces

Horizontal shear force in column bases may be resisted by (see Figure 5.29):

- friction between the base plate, grout and concrete footing;
- shear and bending of the anchor bolts;
- a special shear key, for example a block of I-section or T-section or steel pad welded onto the bottom of the base plate;
- direct contact, e.g. achieved by recessing the base plate into the concrete footing.

Figure 5.29 – Column bases in shear

185

In most cases, the shear force can be resisted through friction between the base plate and the grout. The friction depends on the compressive load and on the coefficient of friction. Pre-stressing the anchor bolts will increase the shear force transfer by friction.

Sometimes, for instance in slender buildings, it may happen that due to horizontal forces (wind loading) columns are loaded in tension. In such cases, the horizontal shear force cannot usually be transmitted through friction. If no other provisions are installed (e.g. shear studs), the anchor bolts will have to transmit these shear forces.

Because the grout does not have sufficient strength to resist bearing stresses between the bolt and the grout, considerable bending of the anchor bolts may occur, as indicated in Figure 5.30. The main failure modes are rupture of the anchor bolts (local curvature of the bolt exceeds the ductility of the bolt material), crumbling of the grout, failure (splitting) of the concrete footing and pull-out of the anchor bolt.

Due to the horizontal displacement, not only shear and bending in the bolts will occur, but also the tensile force in the bolts will be increased due to second order effects. The horizontal component of the increasing tensile force gives an extra contribution to the shear resistance and stiffness. The increasing vertical component results in an extra contribution to the transfer of load by friction, and increases resistance and stiffness as well. These factors explain the shape of the load deformation diagram as given in Figure 5.30. The increase of the load continues until fracture occurs in one of the components of the connection. If the connection is well designed and executed, such fracture will occur at very large deformation (much larger than acceptable in serviceability and ultimate limit states).

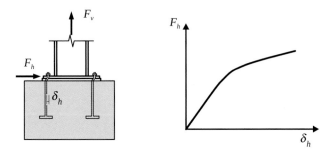

Figure 5.30 – Column base loaded by shear and tension force

The thickness of the grout layer has an important influence on the horizontal deformations. In the tests carried out at the Stevin Laboratory (Bouwman *et al*, 1989), the deformations at rupture of the anchor bolts were between about 15 and 30 mm, whilst grout layers had a thickness of 15, 30 and 60 mm. The deformations have to be taken into account in the check of the SLS. Because of the rather large deformations that may occur, this check may govern the design. The size of the holes may have a considerable influence on the horizontal deformations, especially when oversized holes are applied. In such cases it may be useful to apply larger washers under the nuts, to be welded onto the base plate after erection, or to fill the hole an appropriate resin. For the application of such resin, reference is made to the ECCS recommendations (ECCS, 1994) and EN 1993-1-8.

In Bouwman *et al* (1989), a model for the load deformation behaviour of base plates loaded by combinations of normal force and shear has been developed. For the design of fasteners, the CEB has published a Design Guide (CEB, 1996). The CEB model has been compared with the Stevin Laboratory model and the tests. It appears that the CEB model gives very conservative results, especially when large tensile forces are present and/or the thickness of the grout layer is large. The main reason is that the CEB model does not take account of the positive influence of the grout layer. A summary of the design rules which were proposed in Gresnigt *et al* (2008) for inclusion in EN 1993-1-8 and based on the Stevin Laboratory model is presented below.

It is noted that in steel construction, usually only the steel part of the base plate connection is considered. In the CEB Design Guide much attention is paid to the concrete part. It is recommended to the steel designer to acquire knowledge about the requirements to the reinforced concrete. For detailed design guidance and the various failure modes that may occur in the concrete, reference is made to the CEB Design Guide (CEB, 1996).

The design shear resistance $F_{v,Rd}$ of column base plate joints may be derived as follows:

$$F_{v,Rd} = F_{f,Rd} + nF_{vb,Rd} \qquad (5.31)$$

where:

$F_{f,Rd}$ is the design friction resistance between base plate and grout layer: $F_{f,Rd} = C_{f,d}N_{c,Ed}$

$C_{f,d}$ is the coefficient of friction between base plate and grout layer. The following values may be used:
- for sand-cement mortar: $C_{f,d} = 0.20$
- for special grout: $C_{f,d} = 0.30$

$N_{c,Ed}$ is the design value of the normal compressive force in the column. If the normal force in the column is a tensile force $F_{f,Rd} = 0$

Note: Also, the preload in the anchor bolts contributes to the friction resistance. However, because of its uncertainty (e.g. relaxation and interaction with the column normal force), it is suggested to neglect this action.

n is   the number of anchor bolts in the base plate

$F_{vb,Rd}$ is   the smallest of $F_{1,vb,Rd}$ and $F_{2,vb,Rd}$

$F_{1,vb,Rd}$ is   the bearing resistance for the anchor bolt - base plate; the bearing resistance of the concrete is provided by EN 1992

$F_{2,vb,Rd}$ is   the shear resistance of the anchor bolt

$$F_{2,vb,Rd} = \frac{\alpha_{bc} f_{ub} A_s}{\gamma_{M2}}$$

A_s is   the tensile stress area of the bolt or of the anchor bolt

α_b is   a coefficient depending on the yield strength f_{yb} the anchor bolt: $\alpha_b = 0.44 - 0.0003 f_{yb}$

f_{yb} is   the nominal yield strength the anchor bolt, where: $235\,N/mm^2 \leq f_{yb} \leq 640\,N/mm^2$

f_{ub} is   the nominal ultimate strength the anchor bolt, where: $400\,N/mm^2 \leq f_{ub} \leq 800\,N/mm^2$

γ_{M2} is   the partial safety factor

Prerequisites:

In the above equations, it is a prerequisite that:

- The grout layer is of adequate quality (see relevant applicable reference standards);
- The design strength of the anchor – concrete connection is greater than the design rupture strength of the anchor;
- Other failure modes, like splitting of the concrete and pull out of the anchor, are prevented by adequate design and execution of the anchor in the concrete block. Reference is made to: EN 1992-1-1 (CEN, 2004a), CEB publications (CEB, 1994) and (CEB, 1996).

Chapter 6

MOMENT-RESISTING JOINTS

6.1 INTRODUCTION

Moment-resisting joints is a general term used to cover all joints which transfer significant bending moments between the connected members, but also shear and/or axial forces. These joints may be rigid or semi-rigid, in terms of stiffness, and may exhibit a full or a partial strength resistance level.

The evaluation of their mechanical design properties is a key aspect to which several pages of EN 1993-1-8 are devoted. And as already said, it is based on the application of the component method (see section 1.6.1), that requires the mechanical properties of the active components to be computed before the components are assembled, to derive the global response of the whole joint in a specific loading situation.

This chapter first addresses the characterisation of the mechanical response of individual components covered by EN 1993-1-8. Then, the assembly procedure problem is addressed, successively in terms of resistance, stiffness and deformation capacity. Finally, the application of the component approach to specific joints will be presented. The following joint configurations are considered:

- Steel beam-to-column, beam-to-beam and beam splices;
- Steel column splices;
- Column bases.

6.2 COMPONENT CHARACTERISATION

Joint components are subjected to compression, tension or shear force. In these conditions, they first deform elastically, reach their plastic level of resistance and possibly deform plastically in the absence of brittle failure or development of instability phenomena. In EN 1993-1-8, components are characterised by means of three specific design properties:

- The elastic stiffness, Ek_i where k_i is a stiffness coefficient;
- The plastic resistance $F_{i,Rd}$;
- The (plastic) deformation capacity.

In this section, the determination of these three properties is briefly commented, component by component. For all details about the application of the characterisation formulae, the reader is asked to refer directly to the normative documents.

UK Specific Comment

Publication SCI (2013b) summarises the principal differences between BS 5950 and BS EN 1993-1-8 (BSI, 2000 and 2010) in terms of component characterisation.

With reference to Figure 6.6, it is anticipated that the revised BS EN1993-1-8 will include analytical expressions for the definition of the so-called α coefficient.

Rules of thumb for the characterisation of the ductility of some individual components are given in SCI (1995a).

6.2.1 Column web panel in shear in steel or composite joints

a)   Resistance

The shear force $V_{wp,Ed}$ applied to an unstiffened column web panel (see sub-chapter 2.2) has to be compared to the following expression for the design resistance:

$$V_{wp,Rd} = \frac{0.9 f_{y,wc} A_{vc}}{\sqrt{3}\gamma_{M0}}$$

(6.1)

in which:

A_{vc} is the shear area of the column defined in EN 1993-1-1

$f_{y,wc}$ is the design yield strength of steel for the column web

γ_{M0} is the partial safety factor

In view of the simplified modelling in which reference is made to "joints" (see sub-chapter 2.2), the shear resistance has to be represented as a "beam-force" resistance by means of the transformation parameter β, see section 2.2.4.1:

$$F_{wp,Rd} = \frac{V_{wp,Rd}}{\beta} \qquad (6.2)$$

In Eq. (6.1), the 0.9 factor allows for the detrimental effect of the longitudinal stresses acting in the column on the shear resistance of the web panel. In practical cases, this influence is between 5 and 12%. In EN 1993-1-8, a fixed reduction of 10% of the resistance has been selected for sake of simplicity in application.

The validity of this formula, and of all those provided in the present section, is limited to panels with a web slenderness $d_c/t_w \leq 69\varepsilon$ (d_c is the clear depth of the column web, t_w, the web thickness and $\varepsilon = \sqrt{235/f_{y,wc}}$ with $f_{y,wc}$ expressed in N/mm²). Another limitation applies to double-sided beam-to-column joint configurations without diagonal stiffeners on the column web; in such cases, the two beams should have equal depths.

When the resistance of the panel is insufficient, various reinforcement solutions may be considered. In EN 1993-1-8, guidelines are provided for supplementary web plates or diagonal/transverse column web stiffeners.

For slender panels exceeding the "$d_c/t_w \leq 69\varepsilon$" rule (for instance, for built-up profiles), shear instability of the column web panel develops, which transforms the shear panel into a truss-like system providing extra resistance. Appropriate models exist which allows to determine the design resistance of the panel. Such a model is proposed in Jaspart (1997). Similarly, for double-sided beam-to-column joint configurations without diagonal stiffeners on the column web where the two beams would not be of equal depths, information may be found in Jordão (2008).

For a column web panel forming part of a concrete encased steel column, the resistance of the component is increased by the shear resistance of the concrete, see EN 1994-1-1 section 8.4.4.1.

191

b)   Stiffness

The stiffness coefficient of an unstiffened column web panel is provided by the expression:

$$k_1 = \frac{0.38A_{vc}}{\beta z}$$ (6.3)

For a column web panel with appropriate diagonal stiffeners:

$$k_1 = \infty$$ (6.4)

For an unstiffened panel (without concrete encasement) in a joint with a steel contact plate connection, the stiffness coefficient k_1 may be taken as 0.87 times the value given by Eq.(6.3).

For a column web panel being part of a concrete encased steel column, see EN 1994-1-1 section A.2.3.1.

c)   Deformation capacity

The deformation capacity of steel column web panels is recognised as being rather high. Plastic redistribution between the components requiring ductility from the shear panel may therefore be accommodated without any risk. Moreover joints in which the web panel in shear is the relevant failure mode will possess enough ductility to perform a plastic structural frame analysis. No such high deformation capacity under shear loading may be expected from any concrete encasement that is present.

6.2.2   Column web in transverse compression in steel or composite joints

a)   Resistance

The design resistance of a column web panel in transverse compression is given by:

$$F_{c,wc,Rd} = \frac{\omega k_{wc} b_{eff,c,wc} t_{wc} f_{y,wc}}{\gamma_{M0}}$$ (6.5)

but with:

$$F_{c,wc,Rd} \leq \frac{\omega k_{wc} \rho b_{eff,c,wc} t_{wc} f_{y,wc}}{\gamma_{M1}}$$ (6.6)

This formula is fitted to cover any failure mode in transverse compression, including crushing, crippling, and buckling of the web. Only the "column-sway" buckling mode illustrated in Figure 6.1 is not considered. This one has in fact to be prevented by constructional restraints.

Figure 6.1 – "Column sway" buckling mode of an unstiffened web

In its simplest format, Eq. (6.5) expresses the design resistance of the web as that of an equivalent web zone area (effective width $b_{eff,c,wc}$ times the web thickness t_{wc}) subjected to a stress equal to the design strength of the steel material. The value of the effective width $b_{eff,c,wc}$, see Figure 6.2, depends on the connection type (see EN 1993-1-8 section 6.2.6.2, and EN 1994-1-1 section 8.4.3). Two reduction factors are then added:

- The coefficient ω that allows for the possible stress interaction effects with shear in the column web panel;
- The coefficient k_{wc} which, at first sight, appears to be there to account for the possible stress interaction with the maximum longitudinal compressive stress $\sigma_{com,Ed}$ due to axial force and bending moment in the column, at least for $\sigma_{com,Ed} > 0.7 f_{y,wc}$, see Figure 6.3. But in fact, this factor covers the increase risk of column web instability resulting from the application of a high $\sigma_{com,Ed}$ stress in the column. This explains why this coefficient has not to be applied to column webs in transverse tension (see section 6.2.3 below).

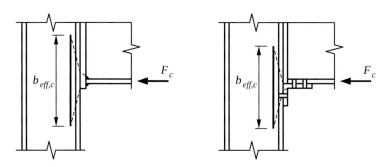

Figure 6.2 – Spread of compression stresses to the column web

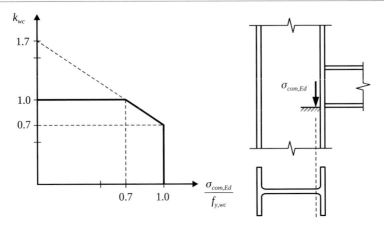

Figure 6.3 – Reduction factor k_{wc}

From a practical point of view, the question has to be raised on how to deal with such interaction factors in the design process. For the ω coefficient, values are given in Table 6.1. For the k_{wc} coefficient, EN 1993-1-8 suggests to consider it as equal to 1.0 as no reduction is necessary in most of the cases. It can therefore be omitted in preliminary calculations when the longitudinal stress is unknown and checked later on.

Table 6.1 – Reduction factor ω for interaction with shear

Transformation parameter β	Reduction factor ω
$0 \le \beta \le 0.5$	$\omega = 1$
$0.5 < \beta < 1$	$\omega = \omega_1 + 2(1 - \beta)(1 - \omega_1)$
$\beta = 1$	$\omega = \omega_1$
$1 < \beta < 2$	$\omega = \omega_1 + (\beta - 1)(\omega_2 - \omega_1)$
$\beta = 2$	$\omega = \omega_2$
$\omega_1 = \dfrac{1}{\sqrt{1 + 1.3\left(b_{eff,c,wc} t_{wc} / A_{vc}\right)^2}}$ and $\omega_2 = \dfrac{1}{\sqrt{1 + 5.2\left(b_{eff,c,wc} t_{wc} / A_{vc}\right)^2}}$	

In Eq. (6.6), an additional resistance limitation is addressed through the ρ coefficient, which allows for the possible risk of instability in the web in the case of slender column sections.

Specific rules to apply to stiffened column webs (transverse stiffeners or supplementary web plates) are provided in EN 1993-1-8.

In the case of concrete-encased columns, an increased level of resistance is reached. EN 1994-1-1 section 8.4.4.2 provides design recommendations to take benefit of this reinforcement.

b)   Stiffness

The stiffness coefficient for the unstiffened column web in transverse compression component is expressed as:

$$k_2 = \frac{0.7 b_{eff,c,wc} t_{wc}}{d_c}$$ (6.7)

For a column web with an appropriate transverse stiffener:

$$k_2 = \infty$$ (6.8)

For an unstiffened column web in a steel contact plate connection, the stiffness coefficient k_2 may be evaluated through a particular expression provided in EN 1994-1-1 section A.2.2.2.

For a concrete-encased steel column, see Eurocode 4 Part 1-8 section A.2.3.2.

c)   Deformation capacity

Little practical information is currently available on the plastic deformation capacity of the column web in transverse compression. In Jaspart (1991), it is specified that a significant deformation capacity may be expected from this component as long as instability does not govern the failure mode. However, to ensure that this component alone would provide enough ductility to the joint is quite questionable. If a plastic global analysis of the structure has to be performed, it is preferable to design the joint by ensuring another failure mode governs, one that exhibits a clear ductile behaviour (for instance, the column web panel in shear, see 6.2.1).

No information is provided in EN 1994-1-1 for concrete-encased steel columns.

6.2.3 Column web in transverse tension

a)   Resistance

The design resistance of the column web in transverse tension is expressed by means of a similar expression to that presented in section 6.2.2:

$$F_{t,wc,Rd} = \frac{\omega \, b_{eff,t,wc} t_{wc} f_{y,wc}}{\gamma_{M0}} \tag{6.9}$$

In the absence of potential instability phenomena, the coefficients k_{wc} and ρ need not be considered. The factor ω, on the other hand, is kept in the formulation so as to allow for possible stress interaction effects with shear in the column web panel.

The main difference between the compression and tension responses is the definition of the effective width, at least where tension forces are transferred through bolt rows. In compression (or in the tension zone of a welded beam-to-column joint), the effective width results from the direct transfer of the forces through the end plate, the column flange, the fillet welds , etc. In the tension zone of a bolted connection, the diffusion of the forces is less direct. In this context, the concept of "equivalent T-stub" described in section 3.5.2 is of great help as it allows defining the effective width $b_{eff,t,wc}$ as equal to the equivalent length of the T-stub, l_{eff}, which models the column flange (see section 6.2.4). According to this, $b_{eff,t,wc}$ is clearly seen as the part of the column web which may be devoted to the transfer of the tension forces by the bolts through the column flange.

Here again, specific rules to apply to stiffened column webs (transverse stiffeners or supplementary web plates) are provided in EN 1993-1-8.

b)   Stiffness

The stiffness coefficient for the unstiffened column web in transverse tension component is expressed as:

$$k_3 = \frac{0.7 b_{eff,t,wc} t_{wc}}{d_c} \tag{6.10}$$

In a bolted connection, this expression has to be applied at the level of individual bolt rows and of groups of bolt rows in tension and $b_{eff,t,wc}$ should be taken equal, for each bolt row (whether it is considered individually or as part of a group), to the smallest of the possible effective lengths (individually or as part of a group of bolt rows).

For a column web with an appropriate transverse stiffener:

$$k_3 = \infty \tag{6.11}$$

c) Deformation capacity

Significant ductility in tension may be expected from normalised steels and so no limitation in plastic deformation capacity may be assumed from this component.

6.2.4 Column flange in transverse bending

a) Resistance

Two different situations can be distinguished to determine the design resistance of a column flange in transverse bending: (i) the connected member is welded to the column flange or (ii) the connected member is bolted to the column flange. In both cases, the column flange can be unstiffened or stiffened. The loading of the component *column flange in transverse bending* is a force in the connected member, perpendicular to the column flange, resulting in bending of the column flange itself.

197

– Column flange in a welded connection

In a welded joint, the resistance $F_{fc,Rd}$ of an *unstiffened* column flange is determined as follows:

$$F_{fc,Rd} = \frac{b_{eff,b,fc}\, t_{fb}\, f_{y,fc}}{\gamma_{M0}} \tag{6.12}$$

whereby:

$b_{eff,b,fc}$ is   the   effective   width   for   bending   of   the unstiffened column flange, $b_{eff,b,fc} = t_{wc} + 2s + 7kt_{wc}$

where:

$s = r_c$ for a rolled I or H section

$$s = \sqrt{2}\, a_c \qquad \text{for a welded I or H section}$$

$$k = \frac{t_{fc}\, f_{y,fc}}{t_{fb}\, f_{y,fb}} \quad \text{but} \quad k \leq 1.0$$

From Eq. (6.12) it can be seen that, in reality, the resistance $F_{fc,Rd}$ of an *unstiffened* column flange is the (plastic) resistance of the beam flange, reduced to account for the unequal stress distribution due to the weaker *unstiffened* column flange. Further explanations and details can be found in section 4.4.2.2.

Even if $b_{eff,b,fc} \leq b_{fb}$, where b_{fb} is the width of the beam flange, the welds connecting the beam flange to the column flange should be designed to resist a force equal to the plastic resistance of the beam flange $b_{fb}\, t_{fb}\, f_{y,fb}/\gamma_{M0}$, i.e. full strength welds are required.

Note that the design resistance of an unstiffened column flange in a welded connection is the same for both an applied tension force and a compression force.

If $b_{eff,b,fc} \leq b_{fb}\, f_{y,fb}/f_{u,b}$, the column flange should be stiffened. If the following conditions for transverse stiffeners in the column web are satisfied, it may be assumed that the resistance of the component *column flange in transverse bending* does not need to be checked:

- the stiffeners are aligned with the beam flange;
- the steel grade of the stiffeners is not lower than that of the beam flange;
- the thickness of the stiffeners is not smaller than the thickness of the beam flange;
- the outstand of the stiffeners is not less than $(b_{fb}-t_{wc})/2$.

– Unstiffened column flange in a bolted connection

Design methods for the so-called equivalent T-stub are presented in section 3.5.2. Note that the T-stub in section 3.5.2 is a T-stub with two bolts per row. However, in practice, also configurations with four bolts per row are used. An extension of the design methods for a T-stub with four bolts per row can be found in Demonceau *et al* (2011). The resistance and failure mode of an *unstiffened column flange in bending*, together with the associated *bolts in tension*, should be taken as similar to those of an equivalent T-stub flange, see Figure 6.4, for both:

- each individual bolt row required to resist tension;
- each group of bolt rows required to resist tension.

Figure 6.4 – Equivalent T-stub flange representing a column flange in bending

The dimensions e_{min} and m for use in 3.5.2 should be determined from Figure 3.30.

The effective length of the equivalent T-stub flange l_{eff} should be determined for the individual bolt rows and the bolt-group as explained in section 3.5.2 using the values given in Table 6.2. Note that the figures in Table 6.2 illustrating the yield line patterns for the "end bolt row" only show cases not yet shown for the "inner bolt row".

– Stiffened column flange in a bolted connection

To increase the resistance of the column flange in bending in a bolted connection, transverse stiffeners and/or appropriate arrangements of diagonal stiffeners may be used. The design resistance is determined in the same way as described in the section before for an *unstiffened column flange*. However, a stiffener welded to the column flange will split possible groups of bolt rows, i.e. the groups of bolt rows on either side of a stiffener should be modelled as separate equivalent T-stubs, see Figure 6.5. The design resistance and the failure modes should be determined separately for each equivalent T-stub.

The effective length of equivalent T-stub flanges l_{eff} should be determined for the individual bolt rows and the bolt-group as explained in section 3.5.2 from the values given for each bolt row in Table 6.2. For those bolt rows adjacent to a stiffener effective length values are given in Table 6.3. Note that the figures in Table 6.3 illustrating the yield line patterns only show cases not yet shown in Table 6.2.

199

Table 6.2 – Effective lengths for an unstiffened column flange in a bolted connection

Bolt row location	Bolt row considered individually		Bolt row considered as part of a group of bolt rows	
	Circular patterns $l_{eff,cp}$	Non-circular patterns $l_{eff,nc}$	Circular patterns $l_{eff,cp}$	Non-circular patterns $l_{eff,nc}$
Inner bolt row	$2\pi m$	$4m + 1.25e$	$2p$	p
End bolt row	The smaller of: $2\pi m$ $\pi m + 2e_1$	The smaller of: $4m + 1.25e$ $2m + 0.625e + e_1$	The smaller of: $\pi m + p$ $2e_1 + p$	The smaller of: $2m + 0.625e + 0.5p$ $e_1 + 0.5p$
Mode 1	$l_{eff,1} = \min(l_{eff,cp}; l_{eff,nc})$		$\Sigma l_{eff,1} = \min(\Sigma l_{eff,cp}; \Sigma l_{eff,nc})$	
Mode 2	$l_{eff,2} = l_{eff,nc}$		$\Sigma l_{eff,2} = \Sigma l_{eff,nc}$	

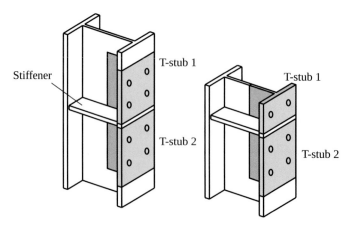

Figure 6.5 – Modelling a stiffened column flange as separate T-stubs

Table 6.3 – Effective lengths for a stiffened column flange in a bolted connection

Bolt row location	Bolt row considered individually		Bolt row considered as part of a group of bolt rows	
	Circular patterns $l_{eff,cp}$	Non-circular patterns $l_{eff,nc}$	Circular patterns $l_{eff,cp}$	Non-circular patterns $l_{eff,nc}$
Bolt row adjacent to a stiffener	$2\pi m$	αm 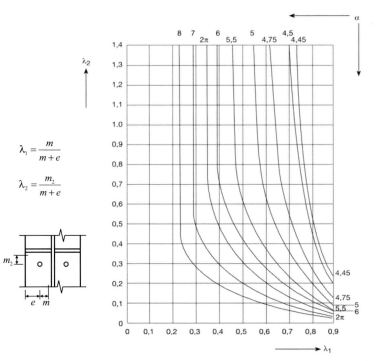	$\pi m + p$	$0.5p + \alpha m - (2m + 0.625e)$
End bolt row adjacent to a stiffener	The smaller of: $2\pi m$ $\pi m + 2e_1$	The smaller of: $e_1 + \alpha m - (2m + 0.625e)$	not relevant	not relevant
Values for α should be obtained from Figure 6.6 For *inner bolt rows* and *end bolt row* not adjacent to a stiffener, see Table 6.2				

201

$$\lambda_1 = \frac{m}{m+e}$$

$$\lambda_2 = \frac{m_2}{m+e}$$

Figure 6.6 – Values for α for effective length of bolt rows adjacent to a stiffener

b)   Stiffness

In bolted connections or in unstiffened welded connections, the stiffness coefficient for the component column flange in transverse bending is expressed as:

$$k_4 = \frac{0.9 l_{eff} t_{fc}^3}{m^3}$$
(6.13)

In a bolted connection, this expression has to be applied at the level of individual bolt rows and of groups of bolt rows in tension and l_{eff} should be taken as equal, for each bolt row (whether it is considered individually or as part of a group), to the smallest of the possible effective lengths (individually or as part of a group of bolt rows) given in Tables 6.2 and 6.3 appropriate. The distance m is defined in Figure 3.29.

For a welded connection, unstiffened or stiffened with appropriate transverse stiffeners:

$$k_4 = \infty$$
(6.14)

c)   Deformation capacity

In bolted connections, significant ductility may be assumed for the column flange in transverse bending if yielding of the flange will be the governing failure. It may be assumed that this component has even sufficient rotation capacity for a plastic analysis, if the design resistance of the joint is governed by the resistance of the column flange in transverse bending and the thickness t_{fc} of the column flange satisfies the following condition:

$$t_{fc} \leq 0.36\, d \sqrt{\frac{f_{ub}}{f_{y,fc}}}$$
(6.15)

where:

d is the bolt diameter;

f_{ub} is the ultimate tensile strength of the bolts;

$f_{y,fc}$ is the yield strength of the column flange.

6.2.5 End plate in bending

a)   Resistance

Similar to a column flange in bending in a bolted connection, the design resistance of the component end plate in bending is calculated using the equivalent T-stub model introduced in section 3.5.2, i.e. the resistance and failure mode of an end plate in bending, together with the associated bolts in tension, should be taken as similar to those of an equivalent T-stub flange for both:

– Each individual bolt row required to resist tension;
– Each group of bolt rows required to resist tension.

The groups of bolt rows at each side of any stiffener connected to the end plate should be treated as separate equivalent T-stubs. In extended end plates, the bolt row in the extended part should also be treated as a separate equivalent T-stub, see Figure 6.7. The resistance and failure mode should be determined separately for each equivalent T-stub.

Figure 6.7 – Modelling an extended end plate as separate T-stub

The dimensions e_{min} and m for use in 3.5.2 should be determined from Figure 3.30 for the part of the end plate located between the beam flanges. For the end plate extension e_{min} should be taken as equal to e_x, see Figure 6.7.

The effective lengths of equivalent T-stub flange l_{eff} should be determined in accordance with section 3.5.2. Values are given for:

- The bolt row outside tension flange of beam in Table 6.4;
- The first bolt row below tension flange of beam in Table 6.3 (bolt row adjacent to a stiffener);
- Other inner bolt rows or other end bolt rows in Table 6.2

The values of m_x, w, e and e_x for use in Table 6.4 should be obtained from Figure 6.7.

Table 6.4 – Effective lengths for an end plate

Bolt row location[*)]	Bolt row considered individually		Bolt row considered as part of a group of bolt rows	
	Circular patterns $l_{eff,cp}$	Non-circular patterns $l_{eff,nc}$	Circular patterns $l_{eff,cp}$	Non-circular patterns $l_{eff,nc}$
Bolt row outside tension flange of beam	The smaller of: $2\pi m_x$ $\pi m_x + w$ $\pi m_x + 2e$	The smaller of: $4m_x + 1.25e_x$ $2m_x + 0.625e_x + e$ $0.5b_p$ $2m_x + 0.625e_x + 0.5w$	not relevant	not relevant
Mode 1	$l_{eff,1} = \min(l_{eff,cp}; l_{eff,nc})$		$\Sigma l_{eff,1} = \min(\Sigma l_{eff,cp}; \Sigma l_{eff,nc})$	
Mode 2	$l_{eff,2} = l_{eff,nc}$		$\Sigma l_{eff,2} = \Sigma l_{eff,nc}$	

[*)] For the first bolt row below tension flange of beam, see *Bolt row adjacent to a stiffener* in Table 6.3.
[*)] For other inner bolt rows or other end bolt rows, see Table 6.2

b) Stiffness

The stiffness coefficient for the component end plate in bending is expressed for a single bolt row in tension as:

$$k_5 = \frac{0.9 l_{eff} t_p^3}{m^3} \qquad (6.16)$$

This expression has to be applied at the level of individual bolt rows and of groups of bolt rows in tension, and l_{eff} should be taken as equal, for each bolt row (whether it is considered individually or as part of a group), to the smallest of the possible effective lengths (individually or as part of a group of bolt rows) given in Tables 6.2, 6.3 and 6.4, as appropriate. m is generally defined in Figure 3.29, but for a bolt row located in the extended part of an extended end plate, $m = m_x$ where m_x is as defined in Figure 6.7.

c)   Deformation capacity

For this component, significant ductility may be assumed if yielding of the end plate will be the governing failure. It may be assumed that an end plate in bending has even sufficient rotation capacity for a plastic analysis, if the design resistance of the joint is governed by the resistance of the end plate in bending and the thickness t_p of end plate satisfy the following condition:

$$t_p \leq 0.36\, d\, \sqrt{\frac{f_{ub}}{f_{y,p}}}\qquad\qquad (6.17)$$

6.2.6 Flange cleat in bending

a)   Resistance

The design resistance of a bolted flange cleat in bending is determined in a very similar way than that of the extended part of an extended end plate, see section 6.2.5a). The resistance and failure mode of an angle cleat in bending and the associated bolts in tension, should be taken as similar to those of an equivalent T-stub flange. The effective length l_{eff} of the equivalent T-stub flange should be taken as $0.5b_a$ where b_a is the length of the angle cleat. The dimensions e_{min} and m for use in section 3.5.2 should be determined from Figure 6.8. Note that, depending on the gap g between the beam end and the column flange, the plastic hinge in the angle cleat may form either in the leg connected to the column flange or in the leg connected to the beam flange, see Figure 6.8.

a) Gap $g = 0.4\, t_a$

b) Gap $g = 0.4\, t_a$

Figure 6.8 – Influence of the gap between the beam and the column

b)   Stiffness

The stiffness coefficient is determined as follows:

$$k_6 = \frac{0.9\, l_{eff}\, t_a^3}{m^3}$$ (6.18)

where:

l_{eff}	$= 0.5 b_a$;
b_a	is the length of the angle cleat;
t_a	is the thickness of the angle cleat;
m	is defined in Figure 6.8.

c)   Deformation capacity

For the component flange cleat in bending, significant ductility may be assumed if yielding of the angle cleat will be the governing failure. It may be assumed that this component alone has sufficient rotation capacity to justify a plastic analysis, if the design resistance of the joint is governed by the resistance of the flange cleat in bending and the thickness of the angle cleat t_a satisfies the following condition:

$$t_a \leq 0.36\, d \sqrt{\frac{f_{ub}}{f_{y,a}}} \tag{6.19}$$

6.2.7 Beam or column flange and web in compression

a)   Resistance

As a result, for instance, of a bending moment applied to the beam, compression develops in the vicinity of one of the beam flanges and the adjacent part of the web. The possible development of plasticity or buckling of this zone is recognised in the component beam flange and web in compression. As shown in Figure 6.9, the concentrated force F_c may be significantly higher than the compression force F induced by the same beam loading, at a certain distance from the connection. In Figure 6.9, forces F_c and F are applied at the centre of gravity of the beam flange.

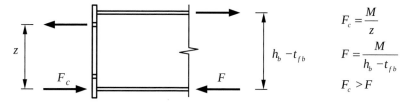

$$F_c = \frac{M}{z}$$

$$F = \frac{M}{h_b - t_{fb}}$$

$$F_c > F$$

Figure 6.9 – Concentrated force F_c and compression force F

This reference is conventionally adopted irrespective of the fact that it does not correspond to the reality, as the compression zone is not only limited to the flange. The localised force F_c is likely to generate premature failure of the zone by plasticity or instability. The verification of this failure mode is achieved by checking that F_c is smaller than or equal to:

$$F_{c,fb,Rd} = \frac{M_{c,Rd}}{h_b - t_{fb}} \qquad (6.20)$$

where:

$M_{c,Rd}$ is the design bending moment resistance of the beam cross section, reduced, if necessary, by the presence of a shear force; this moment takes into account the risk of plate buckling in the compressed part of the web or in the flange;

h_b is the total beam height;

t_{fb} is the beam flange thickness.

The resistance given by Eq. (6.20) combines the resistance of the flange and that of the part of the web in compression. It has to be clearly stated that the formula applies regardless of the connection type (steel bolted, composite, etc.) and of the beam loading (for instance, a beam subjected to compression and bending). In other words, the resistance of the component is "as it is" and the effect of the above-mentioned factors will be covered in the *assembly of components* step of the component method.

b)   Stiffness

In EN 1993-1-8, the stiffness coefficient for is component is taken as equal to:

$$k_7 = \infty \qquad (6.21)$$

This may be because:

- The deformation of the component contributes to that of the beam section, and not to that of the joint; and the deformation of the beam is accounted for by means of the beam stiffness EI_b/L_b, for instance;
- Even if premature yielding of the component leads to localised plastic deformations that are not allowed for through EI_b/L_b, these remain quite limited;
- The reduction of the actual dimensions of the joint to one point located at the intersection between the connected members results, in the frame analysis, to the consideration of higher beam lengths than in reality. The extra deformations of the fictitious beam stubs along the length of the end joints compensates the localised plastic deformations addressed in the previous bullet point.

c) Deformation capacity

The ductility of this component may be estimated through consideration of the beam cross section class which has to be achieved to define the value of $M_{c,Rd}$ in Eq. (6.20).

6.2.8 Beam web in tension

a) Resistance

The similarity between the two components beam web in tension and column web in transverse tension is so obvious that similar formulae are used to derive their design resistance. For the beam web in tension, the expression is as follows:

$$F_{t,wb,Rd} = \frac{b_{eff,t,wb} t_{wb} f_{y,wb}}{\gamma_{M0}} \qquad (6.22)$$

The only difference with Eq. (6.9), besides the replacement of the subscript "c" with "b", is the absence of the reduction factor ω. This absence is justified by the far lower risk of stress interaction with shear stresses in the beam web (see section 6.3.5.2 showing how shear stresses acting in the beam section close to the connection may be distributed, so avoiding interactions).

Here the effective width $b_{eff,t,wb}$ is taken as equal to the equivalent length of the T-stub (l_{eff}) which may be substituted for the end plate in bending obtained for an individual bolt row or a bolt-group, see section 6.2.5a).

b) Stiffness

The stiffness coefficient for the beam web in tension component is expressed as:

$$k_8 = \infty \qquad (6.23)$$

As for the component column or beam flange and web in compression, the deformation of the beam web in tension is assumed to be included, in view of the frame analysis, in the deformation of the beam element and has therefore not to be taken into account a second time through the definition of a finite value of the k_8 stiffness coefficient.

c)   Deformation capacity

Significant ductility in tension may be expected from normalised steels and so no limitation in plastic deformation capacity may be assumed from this component.

6.2.9 Plate in tension or compression

a)   Resistance

The resistance of a plate in tension may result from the yielding of its transverse cross section or from the failure of its net section (cross sectional area reduced by bolt holes or any other openings). Related design formulae are provided in section 6.2 of EN 1993-1-1.

When compression is applied to a plate, the effect of fastening holes is generally disregarded as long as these ones are filled by fasteners, except for oversize and slotted holes. So the cross section check reduces to the verification of the yielding of the plate, again according to section 6.2 of EN 1993-1-1. Plate buckling is practically prevented by an adequate connection detailing (for instance, maximum distance between connecting bolts).

b)   Stiffness

The stiffness coefficient k_9 is assumed to be infinite, for the same reasons as those explained for other components such as beam flange and web in compression, beam web in tension, etc.

c)   Deformation capacity

The ductile behaviour of a plate in tension or compression may only result from the yielding of the cross section. From this point of view, the premature failure in the net cross section of the plate subjected to tension has to be avoided through a proper selection of the layout and dimensions of the plate and bolt holes with the objective to make sure that the plastic resistance of the gross-section is lower than the more brittle failure load in the net-section, as explained in sub-chapter 3.2 of the ECCS Design Manual on "Design of Steel Structures" (Simões da Silva *et al*, 2010).

6.2.10 Bolts in tension

a)   Resistance

Design rules for bolts in tension have been presented in section 3.4.2. In fact, they act in joints as parts of T-stub connections simulating the response of column flanges, cleats or end plate in bending. Related design formulae are provided in section 3.5.2.

b)   Stiffness

A bolt in tension may be idealised as a beam element in tension and its stiffness coefficient can be evaluated, as an approximation, as equal to A_s/L_b where A_s and L_b are respectively defined as the tensile stress area of the bolt and the bolt elongation length, taken as equal to the grip length (total thickness of materials and washers) plus half the sum of the height of the bolt head and the height of the nut.

However, in EN 1993-1-8, another expression of the stiffness coefficient k_{10} is given:

$$k_{10} = \frac{1.6 A_s}{L_b}$$ (6.24)

The factor "1.6" may be justified as follows:

211

- EN 1993-1-8 refers to a "bolt-group", and not to a single bolt, what result in the application of a factor "2.0" (in fact, it should read "a bolt-group with two bolts in tension", in EN 1993-1-8);
- As a result of the development of prying forces in the T-stub, the bolts are overloaded by a factor, derived from the "T-stub theory" (Jaspart, 1997), equal to 1.25, so decreasing their stiffness by $1/1.25 = 0.8$.

For sure, in the case of a "Mode 3" T-stub failure, or a "Mode 1 without prying" failure mode (see section 3.5.2), the application of the "0.8" factor is not relevant and a value of $k_{10} = 2.0 A_s/L_b$ should not be expected. On the other hand, in the two here-above mentioned situations, globally, the stiffness of the bolts remains quite high in comparison of that of the T-stub flange and the "error" remains quite limited. This is why, for sake of simplicity, the "1.6" factor is always selected.

Finally, it has to be mentioned that EN 1993-1-8 does not make any distinction between the stiffness of a non-preloaded or preloaded bolts, even if the use of preloading may result in an increase of the joint stiffness, as shown in Jaspart (1991) where distinct stiffness coefficients are suggested.

c)   Deformation capacity

It is definitely not acceptable to rely on the ductility of bolts in tension. Even if not strictly forbidden, the design of joints which would reach their design resistance through bolt failure should never be recommended. Moreover, this failure mode is prohibited as soon as the designer wants to profit from plasticity. EN 1993-1-8 limits the potential risk of brittle failure through various clauses. For instance:

   – Eq. (6.32) in section 6.4.2 that limits the thickness of the connected plates in comparison to the diameter of the bolts; this formula, see background in (Jaspart, 1997), has to be fulfilled when a plastic global analysis of the frame involving the development of plastic hinges in the joints is targeted;
   – EN 1993-1-8 clause 6.2.7.2(9) where the "$1.9F_{t,Rd}$" limitation, see section 6.3.1, Eq. (6.35), is just of way to limit the redistribution of internal forces in a joint when a risk of premature failure of a bolt row is likely to occur.

212

6.2.11   Bolts in shear

a)   Resistance

The shear resistance of a bolt (and also of a bolt group) has been defined in chapter 3, in the case of prestressed and non-prestressed bolts.

b)   Stiffness

The stiffness coefficient k_{11} relates, in EN 1993-1-8, to a bolt-group with two bolts, sheared along one plane only, see background (Jaspart, 1991):

- for non-preloaded bolts: $$k_{11} = \frac{16 n_b d^2 f_{ub}}{E d_{M16}}$$ (6.25)

- for preloaded bolts: $k_{11} = \infty$ (6.26)

where n_b corresponds to the number of bolt rows, the bolt diameter, and, d_{M16} is the nominal diameter of a M16 bolt.

c)   Deformation capacity

The ductility of a bolt in shear is quite limited and should therefore not to be relied on. This aspect has been already addressed in section 3.5.1.2.

6.2.12   Bolts in bearing (on beam flange, column flange, end plate or cleat)

a)   Resistance

The bearing of the bolts against the plates is ruled by specific design formulae which have been presented in section 3.4.5.

b)   Stiffness

As for bolts in shear, the stiffness coefficient k_{12} corresponds to the deformation in one of the connected plates induced by a bolt-group with two bolts:

- for non-preloaded bolts $$k_{12} = \frac{24 n_b k_b k_t d f_{u,p}}{E} \qquad (6.27)$$

- for preloaded bolts $$k_{12} = \infty \qquad (6.28)$$

where k_b and k_t depend on the connection detailing (see EN 1993-1-8).

c)   Deformation capacity

The bearing deformation may be assumed to be quite ductile, as previously mentioned (section 3.5.1.2).

6.2.13   Concrete in compression including grout

a)   Resistance

This component has been described in section 5.3.3.1, in terms of resistance.

b)   Stiffness

The stiffness coefficient k_{13} is given as (Steenhuis *et al*, 2008):

$$k_{13} = \frac{E_c \sqrt{b_{eff} l_{eff}}}{1.275E} \tag{6.29}$$

where E_c is the concrete Young's modulus, and b_{eff} and l_{eff} are the effective width and length of the T-stub flange, see section 5.3.3.1.

c)   Deformation capacity

In the absence of robust information concerning the ductility of this component, it is advisable not to rely on it.

6.2.14 Base plate in bending under compression

The component which has been presented in section 5.3.3.1 should be named *concrete and base plate under compression*. And, in fact, the information provided in the above section 6.2.13 covers the response of both components. The design resistance and the stiffness coefficient k_{13} is therefore defined as infinite in EN 1993-1-8.

6.2.15 Base plate in bending under tension

a)   Resistance

This component is modelled as a T-stub and therefore the model presented in section 3.5.2 again applies. The only specificity of base plates, in comparison to steel-to-steel connected plates, is the higher probability that prying forces do not develop, as a result of the elongation of the anchor bolts. This aspect has been raised in section 3.5.2.4 where relevant appropriate formulae are presented.

b)   Stiffness

Two expressions are recommended here for the stiffness coefficient k_{15}, according to whether or not prying forces develop as a result of the elongation of the anchors:

- when prying forces develop $k_{15} = \dfrac{0.85 l_{eff} t_p^3}{m^3}$ (6.30)

- when prying forces do not develop $k_{15} = \dfrac{0.425 l_{eff} t_p^3}{m^3}$ (6.31)

where l_{eff} and t_p are defined as the effective length of the T-stub and the thickness of base plate thickness. No explanation other than a lack of consistency between Eq. (6.30) and Eqs. (6.13), (6.16) and (6.18) may explain the difference between the coefficients "0.85" and "0.9" in the expressions of k_{15} (with prying) and k_4, k_5 and k_6.

c)   Deformation capacity

Guidance on the evlaution of the deformation capacity of a column flange in bending (section 6.2.4) and an end plate in bending (section 6.2.5).

6.2.16   Anchor bolts in tension

Basically, an anchor behaves as a bolt in tension. And so the properties may be derived in a similar way than what has been described in section 6.2.10. Some particular features need to be addressed:

–   The design tension resistance of the anchor may be taken as equal to the tensile resistance of its threaded part as long as no premature failure occurs by pull-out failure, concrete cone failure or splitting failure of the concrete. This has been addressed in section 5.3.3.2.

–   The stiffness coefficient k_{17} to be used for an anchor is the same as for a bolt (k_{10}) as long as prying forces develop between the base plate and the concrete foundation (see section 6.2.15). If it is not the case, then the factor "1.6" has to be replaced with "2.0" in Eq. (6.24). And, as raised in section 5.3.3.2, an appropriate definition of the "bolt length" has to be adopted for anchors.

215

6.2.17   Anchor bolts in shear

The topic has already been extensively discussed in section 5.3.4. No further information is required here.

6.2.18   Anchor bolts in bearing

As above, reference is here made to section 5.3.4 where relevant information is provided.

6.2.19 Welds

a)   Resistance

The evaluation of the design resistance of welds has been extensively covered in sub-chapter 4.3. So, no additional information is provided here.

b)   Stiffness

The stiffness coefficient for welds, k_{19}, is assumed to be infinite. Welds are therefore not contributing to the global flexibility of the joints.

c)   Deformation capacity

No significant plastic deformation capacity may be expected from welds, as already expressed in section 4.4.1. Weld failure should always be considered as quite brittle and so as an undesirable failure mode.

6.2.20   Haunched beam

a)   Resistance

The design resistance of a beam flange and of the adjacent zone of the web in compression is given by Eq. (6.20). For a haunched beam, a similar formula applies, but in which $M_{c,Rd}$ is now evaluated neglecting the intermediate flange.

If the height of the beam including the haunch exceeds 600 mm, EN 1993-1-8 specifies that the contribution of the beam web to the design compression resistance should be limited to 20%. This point is still subject to discussions in the scientific community, recent tests indicating that this limitation could be relaxed.

If a beam is reinforced with haunches, the latter should be arranged such that:

  - The steel grade of the haunch should match that of the member;
  - The flange size and the web thickness of the haunch should not be less than that of the member;
  - The angle of the haunch flange to the flange of the member should not be greater than 45°.

Moreover, if a beam is reinforced with haunches, the design resistance of beam web in compression should be determined according to section 6.2.2.

In this evaluation, the length of stiff bearing to the beam should be taken as equal to the thickness of the haunch flange parallel to the beam.

b)   Stiffness

As for the component beam or column flange and web in compression, the relevant stiffness coefficient k_{20}, is assumed to be infinite.

c)   Deformation capacity

Similarly to what has been explained for the component beam or column flange and web in compression, the ductility of this component may be estimated through the definition of the beam cross section class which has to be achieved to define the value of $M_{c,Rd}$ in Eq. (6.20).

6.3   ASSEMBLY FOR RESISTANCE

6.3.1 Joints under bending moments

The procedure for strength assembly as suggested in EN 1993-1-8 is aimed at deriving the value of the so-called design resistance of the joint. For the sake of clarity, it is not presented at first in a general way but is illustrated in the particular case of beam splices with flush end plates.

For the connection represented in Figure 6.10, the distribution of internal forces is quite easy to obtain: the compressive force is transferred at the centroid of the beam flange and the tension force, at the level of the upper bolt row. The resistance possibly associated to the lower bolt row is usually neglected as it contributes in a quite modest way to the transfer of bending moment in the joint (small lever arm).

The design resistance of the joint $M_{j,Rd}$ is associated with the design resistance F_{Rd} of the weakest joint component which can be one of the following: the beam flange and web in compression, the beam web in tension, the plate in bending or the bolts in tension. The bending resistance is given by:

$$M_{j,Rd} = F_{Rd}\, z \tag{6.32}$$

where z is the lever arm ($z = h$).

217

Figure 6.10 – Joint with one bolt row in tension

When more than one bolt row has to be considered in the tension zone (Figure 6.11), the distribution of internal forces is more complex.

Figure 6.11 – Joint with more than one bolt row in tension

Let us assume, initially, that the design of the joint results in a particularly thick end plate in comparison to the bolt diameter, see Figure 6.12. In this case, the distribution of internal forces between the different bolt rows is linear according to the distance from the centre of compression. The compression force F_{Rd} which equilibrates the tension forces acts at the level of the centroid of the lower beam flange. For the sake of clarity, it is shown only in Figure 6.12, but not in the subsequent figures.

218

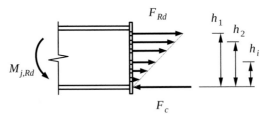

Figure 6.12 – Joint with a thick end plate

The design resistance $M_{j,Rd}$ of the joint is reached as soon as the bolt row subjected to the highest stresses, in reality, the row which is located the farthest from the centre of compression, reaches its design resistance in tension F_{Rd} ($2F_{t,Rd}$ if a Mode 3 failure is assumed in the *plate in bending* component and there are two bolts per row).

As a matter of fact, the quite limited deformation capacity of the bolts in tension does not allow any redistribution of forces to take place between bolt rows.

It is assumed here that the design resistance of the beam flange and web in compression is sufficient to transfer the compression force F_c. The tensile resistance of the beam web is also assumed not to limit the design resistance of the joint. So, $M_{j,Rd}$ is expressed as (Figure 6.12):

$$M_{j,Rd} = \frac{F_{Rd}}{h_i} \sum h_i^2 \qquad (6.33)$$

For thinner end plates, the distribution of internal forces requires more attention. As the moment starts to be increase, forces distribute amongst the bolt rows according to their relative stiffnesse. This stiffness is associated twith that of the part of the end plate adjacent to the bolt row. In the particular case of Figure 6.13, the upper bolt row is characterised by a higher stiffness because of the presence there of the beam flange and web welded to the end plate. Because of the higher stiffness, the upper bolt row is capable of transferring a higher load than the lower bolt rows (Figure 6.13b).

a) Configuration

b) Distribution of the internal forces at the beginning of the loading

Figure 6.13 – Joint with a thin end plate

In EN 1993-1-8, it is assumed that the upper bolt row will reach its design resistance first. This assumption is quite justified in this particular case, but is probably less justified for other connection types, such as for end plate connections with an extended part in the tension zone, where it is usual that the second bolt row just below the flange in tension reaches its design resistance first.

The design resistance of the upper bolt row may be associated to one of the following components: the bolts only, the end plate only, the bolts-plate assembly, or the beam web in tension. If its failure mode is ductile, a redistribution between the bolt rows can take place: as soon as the upper bolt row reaches its design resistance, the supplementary bending moments applied to the joint are carried out by the lower bolt rows (the second, then the third, etc.) which, each one in their turn, may reach their own design resistance.

Failure may occur in three different ways:

a) The plastic redistribution of the internal forces extends to all bolt rows when they have sufficient deformation capacity. The redistribution is said to be "complete" and the resulting distribution of internal forces is called "plastic".

The design moment resistance $M_{j,Rd}$ is expressed as (Figure 6.14):

$$M_{j,Rd} = \sum_i F_{Rd,i} h_i \qquad (6.34)$$

The plastic forces $F_{Rd,i}$ vary from one bolt row to another according to the failure modes (bolts, plate, bolt-plate assembly, beam web, etc.).

EN 1993-1-8 considers that a bolt row possesses sufficient deformation capacity to allow plastic redistribution of internal forces to take place when:

– It is associated to the failure of the beam web in tension or;
– It is associated to the failure of the bolt-plate assembly (including failure of the bolts alone or of the plate alone)

and:

$$F_{Rd,i} \leq 1.9 F_{t,Rd} \qquad (6.35)$$

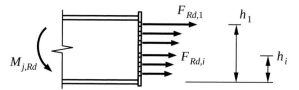

Figure 6.14 – Plastic distribution of forces

b) The plastic redistribution of forces is stopped because of the lack of deformation capacity in the last bolt row which has reached its design resistance ($F_{Rd,i} \leq 1.9F_{t,Rd}$ and linked to the failure of the bolts or of the bolt-plate assembly).

In the bolt rows located lower than bolt row k, the forces are then linearly distributed according to their distance to the point of compression, see Figure 6.15.

The design moment resistance equals:

$$M_{j,Rd} = \sum_{i=1}^{k} F_{Rd,i} + \frac{F_{Rd,k}}{h_k} \sum_{j=k+1}^{n_b} h_j^2 \qquad (6.36)$$

where:

k   is the number of the bolt row of which the deformation capacity is not sufficient.

In this case, the distribution is said "elasto-plastic".

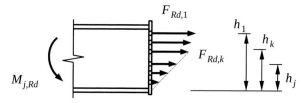

Figure 6.15 – Elasto-plastic distribution of internal forces

c) The plastic or elasto-plastic distribution of internal forces is stopped because the compression force F_c attains the design resistance of the beam flange and web in compression. The moment resistance $M_{j,Rd}$ is evaluated with similar formulae to Eqs. (6.34) and (6.36) in which, obviously, only a limited number of bolt rows are taken into consideration. These bolts rows are such that:

$$\sum_{i=1}^{q} F_i = F_{c,Rd} \tag{6.37}$$

where:

q is   the number of the last bolt row transferring a tensile force;

F_i is   the tensile force in bolt row number i;

$F_{c,Rd}$ is   the design resistance of the beam flange and web in compression.

The application of the above-described principles to beam-to-column joints is quite similar. The design moment resistance $M_{j,Rd}$ is, as for the beam splices, likely to be limited by the resistance of:

- The end plate in bending;
- The bolts in tension;
- The beam web in tension;
- The beam flange and web in compression;

but also by that of:

- The column web in tension;
- The column flange in bending;
- The column web in compression;
- The column web panel in shear.

In the assembly procedure, the possibility to form individual and group yield mechanisms has to be properly considered too. As explained in section 3.5.2.2, various yield mechanisms are likely to form in the connected plates (end plate or column flange) when adjacent bolt rows are subjected to tension forces:

- Individual mechanisms (see Figure 6.16a) which develop when the distance between the bolt rows are sufficiently large;
- Group mechanisms (see Figure 6.16b) including more than one adjacent bolt rows.

Each of these mechanisms are associated to specific design resistances.

When distributing the internal forces, EN 1993-1-8 recommends never to transfer in a given bolt row:

- A higher load than that which can be carried when it is assumed that the considered bolt row is the only one able to transfer tensile forces (individual resistance);
- A load such that the resistance of the whole group to which the bolt row belongs is exceeded.

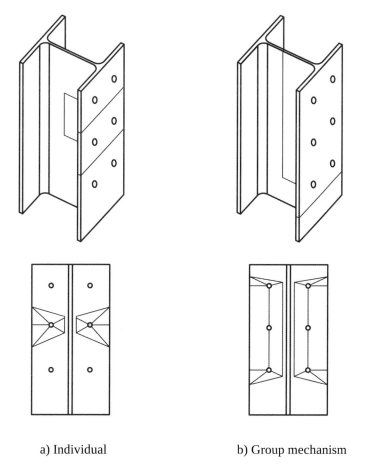

223

a) Individual   b) Group mechanism

Figure 6.16 – Plastic mechanisms

Figure 6.17 gives, in more general terms, a complete overview of the assembly procedure proposed by EN 1993-1-8.

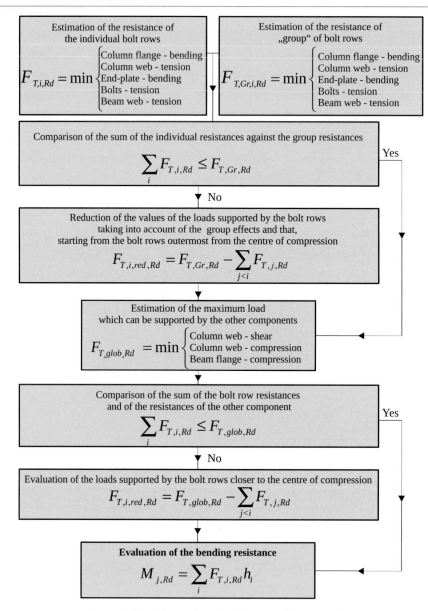

Figure 6.17 – Flow-chart for the assembly procedure

UK Specific Comment

The UK NA to BS EN 1993-1-8 gives specific rules for the full plastic distribution of effective design tension resistance. The procedure is also summarised in SCI (2013a).

6.3.2 Joints under axial forces

The same principles described in the previous section may be applied here. In Figure 6.18a and b, the specific case of a beam splice in compression and tension is illustrated.

a) Beam splice in compression

b) Beam splice in tension

Figure 6.18 – Joint with a symmetrical geometry

The design resistance of this joint under compression equals:

$$N_{j,Rd} = 2N_{c,Rd} \qquad (6.38)$$

where $N_{c,Rd}$ is the design resistance of the beam flange and web in compression component. In a more general case, $N_{j,Rd}$ could be also limited by the resistance of the column web in compression (beam-to-column joint) or by the concrete in compression (column bases).

Under tension forces, the design resistance of the joint is expressed as:

$$N_{j,Rd} = 2N_{t,Rd} \qquad (6.39)$$

where $N_{t,Rd}$ is the smaller of the design resistances of the components end plate in bending and beam web in tension. If more than two bolt rows are present, the design resistance of the joint is evaluated as follows:

$$N_{j,Rd} = \sum N_{ti,Rd} \tag{6.40}$$

where $\Sigma N_{ti,Rd}$ extends to all bolt rows. Here also the possible limitation of the tension resulting from group effects has to be checked.

When the joint detailing is such there is no symmetry with respect to the force axis, the joint is subjected to a combination of axial force and bending moment. This situation is explicitly covered in the next section.

6.3.3 Joints under bending moments and axial forces

6.3.3.1 Introduction

In EN 1993-1-8, it is assumed that the design resistance of a joint in bending is not influenced by the application of an additional axial force $N_{j,Ed}$ as long as the latter remains limited, i.e. lower than 5% of the axial design resistance $N_{pl,Rd}$ of the connected beam (and not of the joint, what is quite surprising as far as the influence of the applied axial load on the joint response is of concern):

$$\left| \frac{N_{j,Ed}}{N_{pl,Rd}} \right| \le 0.05 \tag{6.41}$$

This value seems to be a fully arbitrary and has no scientific background.

However, in many cases, the joints are subjected to significant combined axial loads and bending moments, for instance at the extremities of inclined roof beams or in frames subjected to an exceptional event leading to the loss of a column, situation where significant tying forces can develope in the structural beams above the lost column.

If the criterion given above is not satisfied, EN 1993-1-8 recommends checking the resistance by referring to a $M - N$ interaction diagram defined by the polygon linking the four points corresponding respectively to the hogging and sagging bending resistances in absence of axial forces and to the tension and compression axial resistances in absence of bending. The interaction formulae for normal forces greater than 5% of the axial design resistance $N_{pl,Rd}$ is expressed as follows:

$$\frac{M_{j,Ed}}{M_{j,Rd}} + \frac{N_{j,Ed}}{N_{j,Rd}} \leq 1.0 \qquad (6.42)$$

where:

$M_{j,Rd}$ is the design moment resistance of the joint without the presence of normal forces;

$N_{j,Rd}$ is the design normal resistance of the joint in the absence of bending moments.

EN 1993-1-8 does not say how to derive the tension and compression axial resistances in absence of bending.

The combination of both criteria in Eqs. (6.41) and (6.42) is illustrated in Figure 6.19 by a solid line. In Cerfontaine (2003), a more precise analytical design procedure, based on the component method concept is presented. It has been (i) developed by Cerfontaine to predict the response of ductile and non-ductile steel joints subjected to combined axial loads and bending moments, and (ii) extended to composite joints in Demonceau (2008). This method is briefly introduced in this section.

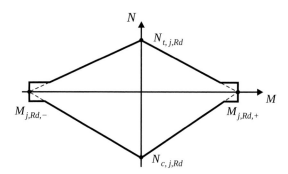

Figure 6.19 – M–N resistant curve for a joint proposed in EN 1993-1-8

227

6.3.3.2 Brief description of the advanced analytical procedure for steel joints

A brief description of the analytical procedure developed by Cerfontaine (2003) is given below. The procedure is in full agreement with the component approach adopted in EN 1993-1-8.

The general concept of the method is first presented. Then, the principles on which it is based to predict the resistance of the joints are given.

a)   General concept

The advanced design method is based on the component approach. As the behaviour of the components is not depending on the type of loading applied to the whole joint, the challenge is to develop an appropriate assembly procedure for bending moments and axial forces. Another aspect to be dealt with is the modification of the list of active components within the joints according to the relative importance of the bending moment and axial load, and obviously according to the respective signs of the applied forces.

Finally, two particular features of the component method have also to be carefully considered within the developed procedure:

– Group effects which may affect the response of *T-stub related* components: column flange in bending, end plate in bending, column web in tension, and the beam web in tension;

– Component interactions that may occur in column components, where three types of stresses interact: shear stresses in the web panel, longitudinal stresses due to axial and bending forces in the column, and transversal stresses due to the load-introduction in the joint area (column web in tension, column web in compression and column web in shear).

Only the procedure to apply to joints activating ductile components at collapse is presented herein. So, the behaviour of each of the constitutive joint components is assumed to be infinitely ductile. As a result, a full plastic redistribution of the internal forces in the joint under bending moment and axial load may be considered.

228

b)   Conventions

Within this section, developments are presented for the case of a bolted end plate steel connection with n_b bolt rows; these ones may potentially, when activated, transfer tension forces; besides that, two compression zones centred at mid-thickness of the upper and lower beam flanges are identified (respectively noted "upper" or "up" and "lower" or "lo"). The compression zones activate two components: the beam flange and web in compression and the column flange in compression.

This leads to a total of $n_b + 2 = n$ rows where internal forces may be transferred from the beam to the column. Conventionally, the tension forces are assumed to be positive (or equal to zero) while a compression force has a negative (or zero) value. All the rows are numbered from 1 to n, starting from the upper row. An example of row numbering is given in Figure 6.20.

Figure 6.20 – Example of row numbering with an extended end plate connection

c)   Equilibrium equations for the connection and load eccentricity

The evaluation of the resistance of the connection based on the static theorem requires, at failure, equilibrium between the distribution of internal forces and the external applied loads. For connections subjected to a bending moment M_{Ed} and an axial load N_{Ed}, the equilibrium criteria are:

$$M_{Ed} = \sum_{i=1}^{n} h_i F_i \text{ and } N_{Ed} = \sum_{i=1}^{n} F_i \tag{6.43}$$

where F_i is the load in row i and h_i is the corresponding lever arm; the latter is defined as the vertical distance between the reference beam point where M_{Ed} and N_{Ed} are computed and the row itself (h_i values are positive for rows located on the upper side of the reference point).

d)   Resistance criteria

According to the static theorem, the resistance of each row, which is equal to the resistance of the weakest component in the row, should never be exceeded. At first sight, it looks easy as long as individual resistances of bolt rows are considered but it is much more questionable when group effects develop in the connections.

Within the developed procedure, any group of rows noted [r; s], i.e. from row r to s, in which group effects appear is considered as an equivalent fictitious row with an equivalent lever arm and a group resistance equal to that of the weakest group component. Therefore, the resistance criteria for each of the rows belonging to the [r; s] group may write, for any constitutive component α:

$$\sum_{i=r}^{s} F_i \leq F_{rs}^{Rd,\alpha} \text{ with } r = 1,...,s \text{ and } s = r, r+1,...,n \qquad (6.44)$$

where $F_{rs}^{Rd,\alpha}$ is the resistance of the component α for the group of r to s. When r is equal to s, $F_{rs}^{Rd,\alpha}$ designates the individual resistance of component α for row r. Such a resistance of the group of rows [r; s] noted F_{rs}^{Rd} is defined as the smallest of the $F_{rs}^{Rd,\alpha}$ values. This situation is illustrated in Figure 6.21 from Cerfontaine (2003) for a connection with three bolt rows (1, 2 and 3), but more generally covers the case of any connection with n rows in which group effects would develop in three bolt rows.

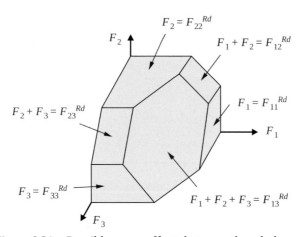

Figure 6.21 – Possible group effects between three bolt rows

e)   Definition of the failure criterion for the whole connection

Details about the application of the static theorem to a connection with n rows are given in Cerfontaine (2003). This application leads to the following definition and presentation of the $M - N$ resistance interaction diagram: the interaction criterion between the bending moment and the axial force at failure is described by a set of $2n$ parallel straight line segments; the slope of each of the $2n$ parallel segments is equal to the value of the lever arm h_k and along these segments, the force F_k in row k varies between 0 at one end and the maximum row resistance at the other end. An example of a $M - N$ resistance interaction diagram is given in Figure 6.22 for a joint with four bolt rows (i.e. $2n = 2 \times (4 + \text{up} + \text{lo}) = 12$ segments); for instance, to pass from point A to point B, row 6 in compression is progressively activated ($F_6 = 0$ at point A where only the rows in tension are activated and $F_6 = F^{Rd,6}$ at point B).

It can be observed in Figure 6.22 (Cerfontaine, 2003) that different values of bending moments correspond to the maximum tensile force; this phenomenon can be easily justified by the evolution of the loads within the rows involved in a group behaviour.

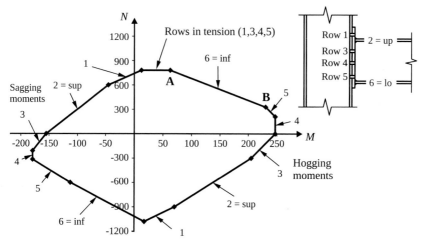

Figure 6.22 – Example of a $M - N$ resistance interaction curve obtained for a four bolt row joint

Each point of the curve is obtained by expressing the quite complex failure criteria given here below for k from 1 to n; this criterion is defined so as to get maximum resistance bending moments by optimising the distributions of the loads amongst the bolt rows with account of the group effects.

Within this expression, it can be observed that two different resistances are attributed to the same row i (F_i^{Rd+} and F_i^{Rd-}). The evaluation procedure of the F_i^{Rd+} and F_i^{Rd-} values is illustrated in Figure 6.23 (Cerfontaine, 2003) for a connection with three bolt rows where the black and white dots respectively show the successive steps for the evaluation of F_i^{Rd+} and F_i^{Rd-}. The objective of the evaluation procedure is to obtain the maximum resistant bending moment by maximising (or minimising if the sign of the bolt row resistance is negative) the forces in the bolt rows which are the most distant from the bolt row k.

$$M = h_k N + \sum_{i=1}^{n} (h_i - h_k) F_i^c$$

$$\text{either} \quad \left. \begin{array}{l} F_i^c = \max(F_i^{Rd+};0) \text{ if } i < k \\ F_i^c = \min(F_i^{Rd+};0) \text{ if } i > k \end{array} \right\} \quad \text{tension in top rows} \qquad (6.45)$$

$$\text{or} \quad \left. \begin{array}{l} F_i^c = \min(F_i^{Rd-};0) \text{ if } i < k \\ F_i^c = \max(F_i^{Rd-};0) \text{ if } i > k \end{array} \right\} \quad \text{tension in bottom rows}$$

with

$$F_i^{Rd+} = \min \left(F_{ri}^{Rd} - \sum_{\substack{j=r \\ i \neq up,lo}}^{i-1} F_j^{Rd+} , r = 1,...,i \right) \qquad \text{for } i < k$$

$$F_i^{Rd+} = F_i^{Rd} \qquad \qquad \qquad \qquad \text{for } i = up,lo > k$$

$$F_i^{Rd-} = \min \left(F_{ir}^{Rd} - \sum_{\substack{j=i+1 \\ i \neq up,lo}}^{r} F_j^{Rd-} , r = i,...,n \right) \qquad \text{for } i > k$$

$$F_i^{Rd-} = F_i^{Rd} \qquad \qquad \qquad \qquad \text{for } i = up,lo < k$$

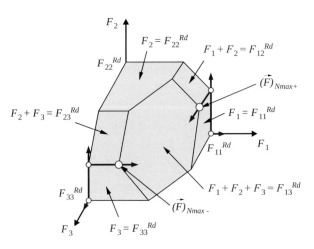

Figure 6.23 – Successive steps for the evaluation of F_i^{Rd+} and F_i^{Rd-} (black and white dots respectively)

The interested reader will find in Cerfontaine (2003) information about the way on how to account for non-ductile components response in the here-above described $M - N$ assembly procedure.

UK Specific Comment

Where a joint is subjected to combined bending moments and axial forces, the concept of an *applied moment modified by axial load* can be used, see SCI (1995a).

6.3.4 $M - N - V$

In end plate connections, the shear forces applied to the joints, in addition to bending moments and/or axial forces, are transferred between the connected members through the bolts, and by bearing of the bolts against the end plate or column flange. Shear failure can then result from a lack of bearing resistance of the connected plates, or from a lack of bolt resistance.

The bearing resistance can be verified as specified in section 3.4.5 while the resistance of a bolt simultaneously subjected to shear and tension forces is expressed as follows (see 3.4.3):

$$\frac{F_{v,Ed}}{F_{v,Rd}} + \frac{F_{t,Ed}}{1.4F_{t,Rd}} \leq 1.0 \tag{6.46}$$

where:

$F_{v,Rd}$ is the design resistance of the bolt in shear;

$F_{t,Rd}$ is the design resistance of the bolt in tension.

It can seen in Eq. (6.46) that the full tension resistance of the bolt $F_{t,Rd}$ may be reached as long as the shear force is not exceeding 0.286 times the shear design resistance $F_{v,Rd}$. Based on this conclusion, the following procedure is suggested in EN 1993-1-8 to check the shear resistance of the joint in presence of bending moments and/or axial forces:

- Identify as an outcome of the $M - N$ assembly procedure the bolt rows which are subjected to tension, n_t, and those bolt rows which are located in the compression zone or which are not activated in tension, n_{nt};
- Consider that the bolt rows which are not subjected to tension may transfer a shear force equal to $2F_{v,Rd}$ (bolt rows with two bolts each);
- Consider that the bolt rows which are subjected to tension may at least transfer a shear force equal to $2 \times 0.286 F_{v,Rd}$ (bolt rows with two bolts each);

  – Evaluate the capacity of the joint to transfer shear forces, in addition to bending moments and/or axial forces, as equal to

$$V_{Rd} = n_t \times 2 \times 0.286\, F_{v,Rd} + n_{nt} \times 2 F_{v,Rd} \tag{6.47}$$

A similar but less safe alternative is to calculate the actual force acting in the n_t bolts subjected tension and evaluating through Eq. (6.46) the actual available resistance of the bolts shear instead of conservatively lowering this value to $0.286 F_{v,Rd}$.

6.3.5 Design of welds

The verification of the welds connecting the beam and the end plate (possibly fitted with haunch or stiffeners) consists in a "global" verification of a particular cross section made of welds only, which may be described as "weld section", see Figure 6.24. In the method defined below a part of the weld section is reserved for the shear force, and does not contribute to the bending or the axial force resistance. Similarly, the weld area that resists the bending moment and the axial force does not contribute to shear resistance.

The verification that is performed for the axial stresses is an elastic verification of the axial stress distribution in the weld section, as no ductility may be considered in welds. The distribution of the axial stresses is based on the Navier assumption. The stresses in the welds are obtained from the internal forces in the frame at the position of the connection.

The procedure described below is appropriate for an E-E or E-P structural design analysis and verification process. For a P-P analysis further recommendations have to be taken into account (see section 6.3.5.3).

6.3.5.1 Definition of the weld section

The weld section must transfer the axial forces and bending moments. As a part of it, a tension zone (top of the connection in Figure 6.24) has to be defined; it is located in the tension part of the joint and comprises the welds which are stressed by the bolt rows, including welds related to stiffeners and haunch, if any. They are assumed to be stressed by the bolt rows because they are located within the effective length obtained from the study of the representative T-stub of the considered bolt row, this length being centred on the bolt row position. As indicated in Figure 6.24, the effective lengths should not be counted twice if they overlap. If there is a gap between successive effective lengths, it should be ignored.

Beam section   Weld section

Figure 6.24 – Definition of the weld section

After the tension part of the weld section has been defined, a part of the web welds is defined as being required to resist the applied shear force. This part of the web welds is arbitrarily located just below the tension area of the weld section.

Tension and shear parts having been defined, the compression area of the weld section can then be identified below the necessary shear area in the remaining available area of the welds. The way in which the longitudinal stresses may distribute in the weld section part which is not reserved for shear is defined below.

6.3.5.2 Position of the neutral axis and calculation of the axial stresses

If an elastic distribution of axial stresses is assumed, the following two equilibrium equations must be respected:

$$N_{Ed} = \int_{WS} \sigma \, d\Omega$$
$$M_{yEd} = \int_{WS} \sigma z \, d\Omega$$

(6.48)

σ is the axial stress at position z and Ω is the weld section. These must be integrated over the weld section, WS.

Two variables are identified: the position of the neutral axis, and the value of the maximum axial tension stress in the weld section. Their definition, through the resolution of the two equations above, is an iterative process for which a practical procedure is suggested below.

Practically speaking, three cases can be met as far as the position of the neutral axis is concerned, see Figure 6.25:

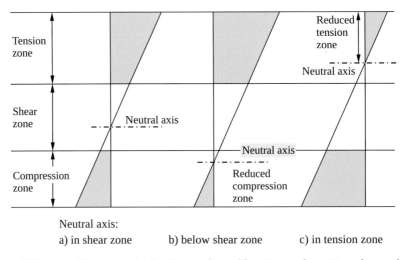

Neutral axis:
a) in shear zone   b) below shear zone   c) in tension zone

Figure 6.25 – Possible stress distributions in the weld section and position of neutral axis

- The neutral axis is located in the shear area (left sketch in Figure 6.25); in this case; the part of the shear area that is below the neutral axis cannot be considered for compression;
- The neutral axis is located below the shear area (sketch in the middle); in this case, the part of the web that is above the neutral axis but below the shear area cannot be used for tension;
- The neutral axis is located in the tension area (right sketch); in this case, the tension zone height must be reduced to account for this specific position of the neutral axis.

As a starting point of the iterative solving process, a position of the neutral axis is guessed (this one changes at each resolution step until convergence is reached). As a result, the height of the compression zone is known and so the weld section is completely defined.

An arbitrary value of the maximum tensile stress σ_t is now guessed – it can even be taken as the yield stress; it allows for the calculation of the corresponding maximum compression stress σ_c. The axial force N and the bending moment about y axis, M_y, corresponding to this assumed situation can thus be calculated. The ratio M_y/N is then evaluated and compared with the same ratio corresponding to the actual internal forces $M_{y,Ed}/N_{Ed}$. If the two ratios are the same, it means that the assumed position of the neutral axis is correct. The actual axial stresses can then be obtained by multiplying the assumed values σ_t and σ_c by the factor $M_{y,Ed}/M_y$ or N_{Ed}/N.

If the ratios M_y/N and $M_{y,Ed}/N_{Ed}$ are not the same, the assumed position of the neutral axis is not correct, and the procedure must be repeated.

6.3.5.3 Design requirements according to the analysis and verification structural design process

The design requirements for joints, as for cross sections, depend on the chosen analysis and verification process. As an example, for a joint in bending (where $M_{j,Rd}$ is the plastic resistance of the joint as given by EN 1993-1-8):

- E-E: $M_{j,Ed} < 2/3 M_{j,Rd}$ no requirement in terms of ductility
- E-P: $M_{j,Ed} < M_{j,Rd}$ ductility is required to redistribute forces inside the joint
- P-P: $M_{j,Ed} < M_{j,Rd}$ sufficient ductility is required to allow plastic redistribution of forces in the structure

In this context, the sizing of the welds has to be addressed accordingly.

a)   E-E and E-P design processes

The design resistance $M_{j,Rd}$ of the joint is evaluated by assuming that the welds are infinitely resistant. The elastic resistance of the joint is then derived (it is conventionally assumed to be equal to $2/3 M_{j,Rd}$).

The resistance of the weld section $M_{w,Rd}$ is evaluated on the basis of the Navier linear distribution.

Finally the resistance of the joint is defined as equal to $\min(2/3 M_{j,Rd}; M_{w,Rd})$ or $\min(M_{j,Rd}; M_{w,Rd})$, when E-E or E-P analysis, respectively, is carried out.

If one wants to fully profit from the joint resistance, i.e. achieve $M_{w,Rd}$ equal to $2/3 M_{j,Rd}$, the welds should be designed such that $M_{w,Rd} \geq 2/3 M_{j,Rd}$.

237

In EN 1993-1-8 clause 6.2.3(4), it is however stated that "in all joints, the sizes of the welds should be such that the design moment resistance $M_{j,Rd}$ of the joint is always limited by the design resistance of its other basic components, and not by the design resistance of the welds". This means that $M_{w,Rd}$ should always be higher than $M_{j,Rd}$. Fundamentally this request appears as unjustified as long as no ductility is requested from the joint in an E-E design process. It is of course conservative and covers the risk of uncertainties on the actual distribution of internal forces in the frame, or the risk of unexpected loading situations resulting from exceptional actions.

b)   P-P design process

The design resistance $M_{j,Rd}$ of the joint is evaluated by assuming that the welds are infinitely resistant.

As the joint has to undergo a plastic rotation and as strain hardening will inevitably occur, resulting in an actual increase of the bending moment in the joint (what is disregarded in a conventional plastic analysis with plastic hinges), the welds, which cannot trigger (brittle) failure, should be calculated, according to EN 1993-1-8, for a bending moment greater than $M_{j,Rd}$.

The resistance of the weld section $M_{w,Rd}$ is evaluated on the basis of the Navier linear distribution; but either $M_{w,Rd} \geq 1.4 M_{j,Rd}$ or $M_{w,Rd} \geq 1.7 M_{j,Rd}$ depending on whether the frame is non-sway or sway, respectively. The condition is more severe for sway frames as higher rotation capacities are required in this case to form a full plastic mechanism.

Finally the resistance of the joint is defined as equal to $M_{j,Rd}$.

238

6.4   ASSEMBLY FOR ROTATIONAL STIFFNESS

6.4.1 Joints under bending moments

6.4.1.1 Refined method

According to the component method, the rotational response of a joint is based on the mechanical properties of its different components. The advantage is that an engineer is able to calculate the mechanical properties of any joint by decomposing the joint into relevant components. Table 6.5 lists the components reported in EN 1993-1-8 and, as an example, identifies the components to be taken into account when calculating the initial stiffness for end plated, welded and flange cleated beam-to-column joints.

Table 6.5 – Overview of components covered by EN 1993-1-8 and list of active components for different joints

Component	Number	End plated	Welded	Flange cleated
Column web panel in shear	1	✓	✓	✓
Column web in compression	2	✓	✓	✓
Column web in tension	3	✓	✓	✓
Column flange in bending	4	✓		✓
End plate in bending	5	✓		
Flange cleat in bending	6			✓
Bolts in tension	10	✓		✓
Bolts in shear	11			✓
Bolts in bearing	12			✓
Concrete in compression	13			
Base plate in bending under compression	14			
Base plate in bending under compression	15			
Anchor bolts in tension	16			

It is assumed that the deformations of the following components: (i) beam flange and web in compression, (ii) beam web in tension and (iii) plate in tension or compression are included in the deformations of the beam in bending. Consequently they are not assumed to contribute to the flexibility of the joint. Haunches, when present, also fall in this category.

The initial stiffness $S_{j,ini}$ is derived from the elastic stiffnesse values of the components. The elastic behaviour of each component is represented by an extensional spring. The force-deformation relationship of this spring is given by:

$$F_i = k_i E \Delta_i \qquad (6.49)$$

239

where:

F_i is the force in the spring i;

k_i is the stiffness coefficient of the component i;

E is the Young's modulus;

Δ_i is the deformation of the spring i.

The spring components in a joint are combined into a spring model. Figure 6.26 shows for example the spring model for an unstiffened welded beam-to-column joint.

Figure 6.26 – Spring model for an unstiffened welded joint

The force in each spring is equal to F. The moment M_{Ed} acting on the spring model is equal to Fz, where z is the distance between the centre of tension (for welded joints, located in the centre of the upper beam flange) and the centre of compression (for welded joints, located in the centre of the lower beam flange). The rotation ϕ in the joint is equal to $(\Delta_1 + \Delta_2 + \Delta_3)/z$. In other words:

$$S_{j,ini} = \frac{M_{Ed}}{\phi} = \frac{Fz}{\dfrac{\sum_i \Delta_i}{z}} = \frac{Fz^2}{\dfrac{F}{E} \cdot \sum_i \dfrac{1}{k_i}} = \frac{Ez^2}{\sum_i \dfrac{1}{k_i}} \qquad (6.50)$$

The same formula applies to end plate joints with a single bolt row in tension, and for flange cleated joints. However, the components to be taken into account are different, see Table 6.5.

Figure 6.27a shows the spring model adopted for end plate joints with two or more bolt rows in tension. It is assumed that the bolt row deformations for all rows are proportional to the distance to the point of compression (located at the centre of the compressed beam flange), but that the elastic forces in each row depend on the stiffness of the components. Figure 6.27b shows how the deformations $k_{i,r}$ of components 3, 4, 5 and 7 are added to an effective spring per bolt row, with an effective stiffness coefficient $k_{eff,r}$ (r, here, is the index of the row number). It is indicated in Figure 6.27c how these effective springs per bolt row are replaced with an equivalent spring acting at a lever arm z. The stiffness coefficient of this effective spring is k_{eq}. The effective stiffness coefficient k_{eq} can directly be applied in Eq. (6.50). The formulae to determine $k_{eff,r}$, z_{eq} and k_{eq} are as follows:

$$k_{eff,r} = \frac{1}{\displaystyle\sum_r \frac{1}{k_{i,r}}}$$ (6.51)

$$z_{eq} = \frac{\displaystyle\sum_r k_{eff,r} h_r^2}{\displaystyle\sum_r k_{eff,r} h_r}$$ (6.52)

$$k_{eq} = \frac{\displaystyle\sum_r k_{eff,r} h_r}{z_{eq}}$$ (6.53)

They can be derived from the sketches of Figure 6.27 by expressing that the moment-rotation response is the same for each of the systems represented in Figure 6.27a-c. An additional condition is that the compressive force in the lower rigid bar is equal in each of these systems.

Figure 6.27 – Spring model for a beam-to-column end plated joint with more than one bolt row in tension

As a result, in the stiffness model used in EN 1993-1-8:

- The internal forces are in equilibrium with the bending moment;
- Compatibility of the displacements is ensured through the assumption of an infinitely rigid transverse stiffness of the beam cross section;
- The plasticity criterion is fulfilled as long as the elastic resistance of the springs is not reached;
- No ductility requirement is likely to limit the deformation capacity of the springs in the elastic range of behaviour.

The respect of these four main requirements (which any distribution of internal forces should satisfy, from a theoretical point of view) allows the model to be considered accurate.

Beyond the elastic resistance ($M_{j,Ed} > 2/3M_{j,Rd}$), the response of the joint becomes non-linear, as depicted in Figure 6.28.

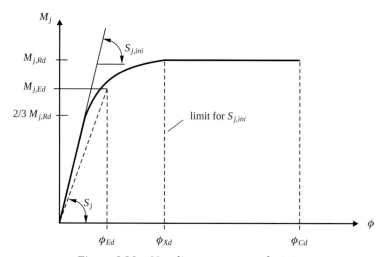

Figure 6.28 – Non-linear response of a joint

In the elastic zone, the joint stiffness is equal to the elastic initial stiffness, $S_{j,ini}$. Beyond the elastic resistance limit, the secant stiffness S_j slowly decreases until $M_{j,Rd}$ is reached. In EN 1993-1-8, the following expression is given for S_j, with reference to a factor $\mu = S_{j,ini}/S_j$:

$$S_j = \frac{1}{\mu}S_{j,ini} = \frac{Ez^2}{\mu\sum_i \frac{1}{k_i}} \tag{6.54}$$

The limit of validity of this formula is ϕ_{Xd}, indicated in Figure 6.28. Values for μ are derived as follows (see also Table 6.6):

$$\mu = 1.0 \quad \text{for } M_{j,Ed} \leq \frac{2}{3} M_{j,Rd} \tag{6.55}$$

$$\mu = \left(\frac{1.5 M_{j,Ed}}{M_{j,Rd}} \right)^{\psi} \quad \text{for } \frac{2}{3} M_{j,Rd} < M_{j,Ed} \leq M_{j,Rd} \tag{6.56}$$

Table 6.6 – Values of the ψ coefficient

Connection type	μ	ψ
Welded	3.0	2.7
Bolted end plates	3.0	2.7
Bolted angle flange cleats	3.5	3.1
Base plate connections	3.0	2.7

Beyond $M_{j,Rd}$, a yield plateau is assumed, but obviously as long as some plastic ductility is available (ductility limit ϕ_{Cd}).

243

6.4.1.2 Simple prediction of the initial stiffness

In the preliminary frame design phase, it is difficult to assess the stiffness of the (semi-rigid) joints; indeed the joints have not been designed yet. To overcome this problem, some simplified formulae have been derived by Steenhuis based on EN 1993-1-8 (Maquoi et al, 1998). By means of these formulae, the designer can determine the stiffness of a joint by selecting the joint configuration only.

Of course these formulae are based on some fixed choices regarding the connection detailing. These are:

- For end plate connections:

  - The connection has two bolt rows only in the tension zone;

- The bolt diameter is approximately 1.5 times the column flange thickness;
- The location of the bolt is as close as possible to the root radius of the column flange, the beam web and flange (about 1.5 times the thickness of the column flange);
- The end plate thickness is similar to the column flange thickness.

– For cleated *connections:*

- The connection has one bolt row in the tension zone;
- The bolt diameter is approximately 1.5 times the cleat thickness;
- The location of the bolt is as close as possible to the root radius of the column flange and the cleat (about 1.5 times the thickness of the column flange);
- The cleat thickness is whenever possible similar to the column flange thickness.

The approximate value $S_{j,app}$ of the initial joint stiffness is expressed as:

$$S_{j,app} = \frac{E\, z^2\, t_{fc}}{C} \tag{6.57}$$

where the values of the factor C are given in Table 6.7 for different joint configurations and loadings. These values of the joint stiffness involve two geometrical parameters only, z and t_{fc}. In an extended end plate connection with two bolt rows, the lever arm z can be approximately taken as equal to the beam depth. For the same joint with a haunch, z is equal to the sum of the beam depth and the haunch depth.

Table 6.7 – Good guess of the initial stiffness for typical beam-to-column joint configurations

Joint configuration			C
1	Extended end plate, single sided, unstiffened ($\beta = 1$)		13

Table 6.7 – Good guess of the initial stiffness for typical beam-to-column joint configurations (continuation)

	Joint configuration		C
2	Extended end plates, double sided, unstiffened, symmetrical ($\beta=0$)		7.5
3	Extended end plate, single sided, stiffened in tension and compression ($\beta=1$)		8.5
4	Extended end plates, double sided stiffened in tension and compression symmetrical ($\beta=0$)		3
5	Extended end plate, single sided, Morris stiffener ($\beta=1$)		3
6	Flush end plate, single sided ($\beta=1$)		14

Table 6.7 – Good guess of the initial stiffness for typical beam-to-column joint configurations (continuation)

Joint configuration			C
7	Flush end plates, double sided, symmetrical ($\beta=0$)		9.5
8	Flush end plate, single sided, cover plate at column top ($\beta=1$)		11.5
9	Flush end plates, double sided, cover plate at column top, symmetrical ($\beta=0$)		6
10	Welded joint, single sided, unstiffened ($\beta=1$)		11.5
11	Welded joints, double sided, unstiffened, symmetrical ($\beta=0$)		6

Table 6.7 – Good guess of the initial stiffness for typical beam-to-column joint configurations (continuation)

Joint configuration			C
12	Cleated, single sided ($\beta = 1$)		70
13	Cleated, double sided, symmetrical ($\beta = 0$)		65

Note: For the rare cases of double-sided joint configurations where $\beta = 2$ (unbalanced moments), the value of the C factor is obtained by adding 11 to the relevant value for symmetrical conditions (balanced moments).

6.4.2 Joints under bending moments and axial forces

For a joint subjected to a combination of axial forces and bending moments, rotational and axial stiffnesses have to be evaluated; these stiffness values depend of the activated components under the considered combined applied loads.

The flexural and extensional stiffnesses are given by the following formulae (Cerfontaine, 2003):

- Stiffness in bending: $K_M = M/\varphi_{el}$
- Stiffness in tension/compression: $K_M = N/\Delta_{el}$

where φ_{el} is the rotation of the joint section and Δ_{el}, the elongation at the reference axis ($h = 0$) (Figure 6.29).

The neutral axis corresponds to the zero-displacement level. Its position is given by $h_0 = -\Delta_{el}/\varphi_{el}$, under the assumption of plane cross section.

247

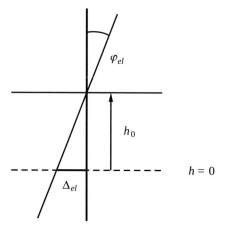

Figure 6.29 – Plane deformation of the joint section

The elastic displacement of any row i is given by:

$$\Delta_{i,el} = \frac{F_{i,el}}{K_{i,ini}} = \Delta_{el} + h_i \varphi_{el} \qquad (6.58)$$

where $K_{i,ini}$ is the initial stiffness of row i, computed as the geometric mean of the initial rigidities of the components in row i.

So, the equilibrium equations become:

$$N_{el} = \sum F_{i,el} = \left(\sum K_{i,ini}\right)\Delta_{el} + \left(\sum K_{i,ini} h_i\right)\varphi_{el} \qquad (6.59)$$

$$M_{el} = \sum F_{i,el} h_i = \left(\sum K_{i,ini} h_i\right)\Delta_{el} + \left(\sum K_{i,ini} h_i^2\right)\varphi_{el} \qquad (6.60)$$

in which only the active rows are taken into account in the summations.

Further developments lead to the following expressions of the position of the neutral axis and the initial rigidities as functions of $e_{el} = M_{el}/N_{el}$:

$$h_{0,el} = \frac{\sum K_{i,ini} h_i \left(h_i - e_{el}\right)}{\sum K_{i,ini} \left(h_i - e_{el}\right)} \qquad (6.61)$$

$$K_{N,el} = \frac{\sum K_{i,ini} \sum K_{i,ini} h_i^2 - \left(\sum K_{i,ini} h_i\right)^2}{\sum K_{i,ini} h_i \left(h_i - e_{el}\right)} \qquad (6.62)$$

$$K_{M,el} = \frac{e_{el}\left[\sum K_{i,ini}\sum K_{i,ini}h_i^2 - \left(\sum K_{i,ini}h_i\right)^2\right]}{\sum K_{i,ini}\left(e_{el} - h_i\right)} = -e_{el}h_{0,el}K_{N,el} \qquad (6.63)$$

As the summations involve only the active rows, which are unknown until the position of the neutral axis has been determined, an iterative procedure is required to find $h_{0,el}$. For a given value of e_{el}, an assumption has first to be made on the value of $h_{0,el}$ and the corresponding active rows. Then, $h_{0,el}$ can be computed using the previous formula and the assumption can be checked. If it is correct, then $K_{N,el}$ and $K_{M,el}$ can be computed. Otherwise, another assumption has to be made until the actual activated rows are found.

UK Specific Comment

Rotational stiffness is not normally evaluated in UK design practice, which usually relies on past experience.

6.5 ASSEMBLY FOR DUCTILITY

Currently EN 1993-1-8 does not provide practitioners with a real "component approach for ductility" in which the ductility of the individual components should first be evaluated before an assembly procedure is used to derive the value of the rotational ductility. Despite some on-going efforts to derive such a "component approach for ductility", it has to be recognised that even today (end of 2016) no consistent set of recommendations may be proposed for inclusion in EN 1993-1-8.

So, as already mentioned in sub-chapter 2.5, the quantification of a maximum rotation value is replaced in EN 1993-1-8 by the proposal of criteria which allow a classification of the joints in three categories:

- *Class 1 joints*: ductile joints exhibiting an adequate rotation capacity for a global plastic analysis with plastic hinges developing in the joints;
- *Class 2 joints*: semi-ductile joints with too limited plastic rotational capacity to achieve a plastic analysis;
- *Class 3 joints*: brittle behaviour requesting an elastic frame analysis combined with an elastic verification of the joint resistance.

249

Before describing these classification criteria, two statements have to be made; they are expressed below.

- If the design resistance of the joint exceeds by at least 20% the design resistance of the weakest of the connected members, it may be assumed that the plastic hinge will not form in the joint, even in the case of material over strength. Therefore, no plastic rotational capacity is required from the joint, but ductility should be provided by the connected members.
- Welds are brittle by nature and so their failure should be strictly avoided if any ductility is required. This results in specific requirements that have been already addressed in section 6.3.5.3.

That said, the classification criteria for bolted and welded steel joints are presented respectively in sections 6.5.1 and 6.5.2. They are valid for S235, S275 and S355 steel grades and for joints in which the design value of the axial force N_{Ed} in the connected member does not exceed 5% of the design plastic resistance $N_{pl,Rd}$ of its cross section (see section 6.3.3.1).

In cases not covered by sections 6.5.1 and 6.5.2, the rotation capacity may be determined by testing in accordance with EN 1990, Annex D (CEN, 2002). Alternatively, appropriate calculation models may be used, provided that they are based on the results of tests in accordance with EN 1990.

UK Specific Comment

Some rules of thumb for the characterisation of the ductility of moment-resisting joints are given in SCI (1995a).

6.5.1 Steel bolted joints

Figure 6.30 presents the flow chart for the ductility classification of bolted joints. Sufficient ductile deformations are possible when the following basic components govern failure:

- Column web in shear;
- Column flange, end plate or cleat in bending.

When the column web panel in shear determines the moment design resistance, the available rotation capacity is assumed to be sufficient for a rigid-plastic analysis, provided that shear buckling of the web does not occur. To respect this condition, the web slenderness has not to exceed a slenderness limit specified in section 6.2.1a.

When the moment resistance of the joint is limited by yielding of the column flange, of the end plate or of the connecting cleats, sufficient rotation capacity for rigid-plastic analysis is also assured, provided that failure is due to pronounced plastic bending deformations of the plate components, and not resulting from bolt failure. Requirements to respect this requirement are indicated in Figure 6.32. These may also be used at preliminary design stages to ensure a ductile response of the joints.

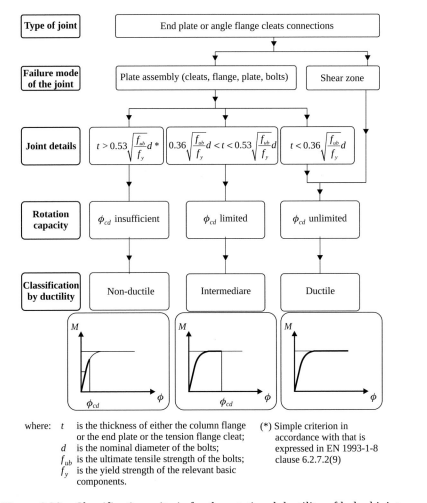

where: t is the thickness of either the column flange or the end plate or the tension flange cleat;
d is the nominal diameter of the bolts;
f_{ub} is the ultimate tensile strength of the bolts;
f_y is the yield strength of the relevant basic components.

(*) Simple criterion in accordance with that is expressed in EN 1993-1-8 clause 6.2.7.2(9)

Figure 6.30 – Classification criteria for the rotational ductility of bolted joints

251

If the moment capacity is limited by the load capacity of the bolts in shear, a sufficient rotation capacity for the rigid-plastic analysis cannot be expected.

The flow chart in Figure 6.30 covers only some of the possible failure modes; for those not covered, the rotation capacity may be determined by testing or through models validated by testing.

6.5.2 Steel welded joints

Figure 6.31 summarises the procedure to classify welded joints for rotational ductility. It is seen that ductile deformations may only be expected from the following basic components:

– Column web in shear;
– Column web in compression;
– Column web in tension.

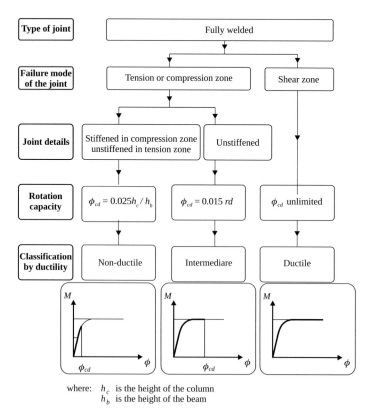

where: h_c is the height of the column
h_b is the height of the beam

Figure 6.31 – Classification criteria for the rotational ductility of welded joints

When the bending resistance of the joint is limited by the resistance of the column web panel in shear, a rigid-plastic analysis involving the development of plastic hinges in the joints may be considered provided that the shear buckling of the web does not occur. As for bolted joints, the web slenderness cannot exceed the slenderness limit specified in section 6.2.1a to comply with this last condition. When the two other above-mentioned failure modes prevail, an intermediate (limited) rotational ductility is available. To know whether it is sufficient for a rigid-plastic structural analysis involving plastic hinges in joints, this available rotation capacity should be compared to the required one.

6.6 APPLICATION TO STEEL BEAM-TO-COLUMN JOINT CONFIGURATIONS

6.6.1 Extended scope

According to the component method approach used in EN 1993-1-8, a joint is considered as a set of individual components and the derivation of its design properties relies on a three-step procedure successively involving (section 1.6.1.2):

– The identification of the basic components;
– The characterisation of the component response;
– The assembly of the components.

Fundamentally, its scope of application is extremely large in terms of covered joint configurations and connection types; but in practice, it is limited by the available knowledge the user has on component characterisation and component assembly, although EN 1993-1-8 provides information on several commonly found components (EN 1993-1-8, Table 1.2 in section 1.6 and section 6.2) and on assembly procedures (EN 1993-1-8, sections 6.3 to 6.5). Its resulting wide scope of application is nevertheless unlikely to satisfy all the designer's needs, in particular when tubular members or innovative/less traditional connection types and connectors are used. Some examples of joints that are not covered by EN 1993-1-8 are shown in Figures 6.32-6.37. In such cases, one often comes to the conclusion that the lack of knowledge relates to the characterisation of the response of some specific "unknown" individual components. In Jaspart *et al* (2005), the authors have gathered most of the scientifically validated component characterisation models that were available in in 2005 and which are not yet included in EN 1993-1-8.

Furthermore the component approach is nowadays a reference for many researchers in the world who regularly develop characterisation models for new components, so multiplying the user's design possibilities. As examples, Jaspart (1997) and Demonceau *et al* (2011) provide information for the design of the joints illustrated in Figures 6.32-6.34.

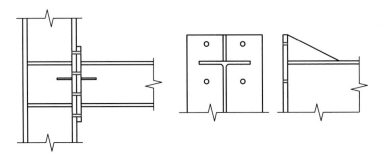

Figure 6.32 – Bolted end plate joints with various reinforcing stiffeners

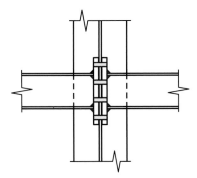

Figure 6.33 – Weak axis beam-to-column joints with H or I members

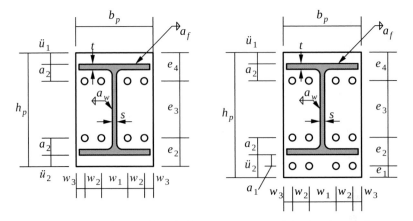

Figure 6.34 – Beam splice with end plate connection exhibiting four bolts per row

Figure 6.35 – Beam-to-column joints with RHS members

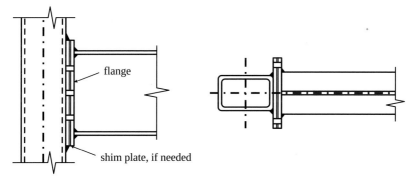

Figure 6.36 – Bolted flange plate connection in a beam-to-column joint with an RHS column

Figure 6.37 – Beam-to-column connections with a continuous RHS beam

UK Specific Comment

With reference to Figure 6.36 it is noted that this joint configuration is hardly (if ever) used in the UK.

6.6.2 Possible design simplifications for end plate connections

EN 1993-1-8 allows the designer to use a simplified procedure for the estimation of the resistance and the stiffness of joints with extended end plates subjected to bending, as explained below.

6.6.2.1 Design moment resistance

As a conservative simplification, the design moment resistance of an extended end plate joint with only two rows of bolts in tension may be estimated as indicated in Figure 6.38, provided that the total design resistance F_{Rd} does not exceed $3.8F_{t,Rd}$, corresponding to the design resistance of the bolts in tension.

Figure 6.38 – Simplified design model for bending resistance

In this case the whole tension region of the end plate may be treated as a single basic component as long as the two bolt rows are approximately equidistant either side of the tension beam flange. $F_{1,Rd}$ is first evaluated and

$F_{2,Rd}$ is then assumed to be equal to $F_{1,Rd}$; finally F_{Rd} may be taken as equal to $2F_{1,Rd}$. The $3.8F_{t,Rd}$ limitation ensures the safe character of the approach by avoiding that the second failure mode exhibits a brittle failure mode before the calculated $M_{j,Rd}$ bending resistance of the whole joint is reached. In report CRIF (1996), this simplification has been followed to derive practical design tools (simplified design procedure for specific joint configurations and tables of standardised joints).

6.6.2.2 Initial stiffness

A similar simplification is suggested for the computation of the initial stiffness. In this case, a set of modified values may be used for the stiffness coefficients of the related basic components to allow for the combined contribution of both bolt rows. Each of these modified values should be taken as twice the corresponding value for a single bolt row in the extended part of the end plate. This approximation leads to a slightly lower estimate of the rotational stiffness.

When using this simplified method, the lever arm z should be taken as equal to the distance from the centre of compression to a point midway between the two bolt rows in tension, see Figure 6.39.

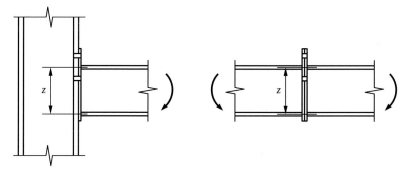

Figure 6.39 – Lever arm for simplified method

6.6.3 Worked example

6.6.3.1 General data

The chosen example is the beam-to-column joint in a typical steel frame shown in Figure 6.40. The detailing of the joint is also shown in this figure. It is an end plate joint, with the plate extended in the tensile zone and four rows of bolts. Steel grade S235 is used.

The objective of this example is to characterise the joint in terms of resistance and stiffness when subjected to hogging moments. The prediction of the resistance in shear will also be addressed.

The partial safety factors (as recommended in EN 1993-1-8) are as follows:

$$\gamma_{M0} = 1.00$$
$$\gamma_{M1} = 1.00$$
$$\gamma_{M2} = 1.25$$

Detailed design of the joint according to the recommendations given in the EN 1993-1-8 is given below. The main steps described in Figure 6.17 will be followed to calculate the resistance in bending.

Figure 6.40 – Joint in a steel building frame and joint detailing

6.6.3.2 Determination of the component properties

a)   Component 1 – Column web panel in shear

According to the catalogue of profiles, the shear area is equal to 4495 mm². So, the resistance of the column web panel in shear, Eq. (6.1), is equal to:

$$V_{wp,Rd} = \frac{0.9 \times 235 \times 4495}{1000\sqrt{3} \times 1.0} = 549 \text{ kN}$$

b)   Component 2 – Column web in compression

The effective width is equal to:

$$b_{eff,c,wc,2} = t_{fb} + 2\sqrt{2}a_{fb} + 5(t_{fc} + s) + s_p$$
$$= 16 + 2\sqrt{2} \times 8 + 5(16.5 + 27) + 15 + 3.69 = 274.8 \text{ mm}$$

In most of the cases, the longitudinal stress in the web of the column $\sigma_{com,Ed} \le 0.7f_{y,wc}$ and, so $k_{wc} = 1$. In the case of a single-sided joint, β can be safely taken as equal to 1. Accordingly:

$$\omega = \omega_1 = \frac{1}{\sqrt{1 + 1.3\left(\dfrac{b_{eff,c,wc}t_{wc}}{A_{vc}}\right)^2}}$$

$$= \frac{1}{\sqrt{1 + 1.3\left(\dfrac{274.81 \times 9.5}{4495}\right)^2}} = 0.83$$

The estimation of the slenderness of the web is used for the definition of the reduction factor ρ.

$$\bar{\lambda}_p = 0.932\sqrt{\frac{b_{eff,c,wc}d_{wc}f_{y,wc}}{Et_{wc}^2}}$$

$$= 0.932\sqrt{\frac{274.81 \times 243 \times 235}{210000 \times 9.5^2}} = 0.85$$

$$\rho = \frac{\left(\overline{\lambda}_p - 0.2\right)}{\overline{\lambda}_p^2}$$

$$= \frac{(0.85 - 0.2)}{0.85^2} = 0.90$$

As $\rho < 1$, the resistance in compression will be limited by a local buckling phenomenon. By using Eqs. (6.5) and (6.6):

$$F_{c,wc,Rd} = \frac{0.83 \times 1.0 \times 274.8 \times 9.5 \times 235}{1.0 \times 1000} = 512 \text{ kN}$$

$$\leq \frac{0.83 \times 1.0 \times 0.90 \times 274.8 \times 9.5 \times 235}{1.0 \times 1000} = 461 \text{ kN}$$

c)   Component 3 – Column web in tension

The values of effective lengths need to be known for the estimation of the resistance of this component. These values are the same as those to be calculated for the component 4 column flange in bending (see section 6.6.3.2d). Four resistances have to be determined see also Figure 6.41:

 – Individual bolt rows;
 – Combined rows 1 and 2;
 – Combined rows 1, 2 and 3;
 – Combined rows 2 and 3.

Figure 6.41 – Bolt rows and groups of bolt rows

In a first step, the behaviour of each individual bolt row is considered. At this level, the resistance of the bolt rows 1, 2 and 3 are identical, see Eq. (6.9):

$$F_{t,wc,Rd} = \frac{0.86 \times 247.1 \times 9.5 \times 235}{1.0 \times 1000} = 474 \text{ kN}$$

where: $b_{eff,t,wc} = l_{eff,b,fc} = 247.1$ mm

$$\omega = \frac{1}{\sqrt{1+1.3\left(\dfrac{247.1\times9.5}{4495}\right)^2}} = 0.86$$

The resistance of the combined bolt rows 1 and 2 is equal to, see Eq. (6.9):

$$F_{t,wc,Rd} = \frac{0.78\times332.1\times9.5\times235}{1.0\times1000} = 579 \text{ kN}$$

where : $b_{eff,t,wc} = l_{eff,b,fc} = (166.05 + 166.05) = 332.1$ mm

$$\omega = \frac{1}{\sqrt{1+1.3\left(\dfrac{332.1\times9.5}{4495}\right)^2}} = 0.78$$

For the combined bolt rows 1, 2 and 3, see Eq. (6.9):

$$F_{t,wc,Rd} = \frac{0.710\times412.1\times9.5\times235}{1.0\times1000} = 653 \text{ kN}$$

where : $b_{eff,c,wc} = l_{eff,b,fc} = (166.05 + 82.5 + 163.55) = 412.1$ mm

$$\omega = \frac{1}{\sqrt{1+1.3\left(\dfrac{412.1\times9.5}{4495}\right)^2}} = 0.71$$

Finally, for the combined bolt rows 2 and 3, see Eq. (6.9):

$$F_{t,wc,Rd} = \frac{0.79\times327.1\times9.5\times235}{1.0\times1000} = 573 \text{ kN}$$

where : $b_{eff,t,wc} = l_{eff,b,fc} = (163.55 + 163.55) = 327.1$ mm

$$\omega = \frac{1}{\sqrt{1+1.3\left(327.1\times9.5\,/\,4495\right)^2}} = 0.79$$

d)   Component 4 – Column flange in bending

For this component, the T-stub model is used. The following geometrical parameters need to be determined:

$$m = \frac{w_1}{2} - \frac{t_{wc}}{2} - 0.8r_c = \frac{120}{2} - \frac{9.5}{2} - 0.8 \times 27 = 33.7 \text{ mm}$$

$$e = \frac{b_c}{2} - \frac{w_1}{2} = \frac{300}{2} - \frac{120}{2} = 90 \text{ mm}$$

$$e_1 = 50 \text{ mm}$$

$$e_{min} = \min(e;w_2) = \min(90;60) = 60 \text{ mm}$$

Table 6.4 of EN 1993-1-8 allows computing the effective lengths corresponding to different patterns of yielding lines and that, for the following cases, see Figure 6.42:

- Individual bolt rows
- Combined bolt rows 1 and 2
- Combined bolt rows 1, 2 and 3
- Combined bolt rows 2 and 3

Group for rows 1+2   Group for rows 2+3   Group for rows 1+2+3

Figure 6.42 – Bolt rows and groups of bolt rows

The contribution of bolt row 4 to the resistance in bending is limited, due to its close position to the centre of compression; accordingly, this contribution is neglected here. In this case bolt row 4 may be assumed to contribute with its full shear resistance to the design shear resistance of the joint. However, if needed, also bolt row 4 may be considered to be in tension zone and hence, its contribution to the moment resistance could be taken into account.

Row 1 – Individual effective lengths:

 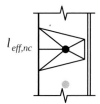

$$l_{eff,cp} = 2\pi m = 2\pi \times 33.65 = 211.4 \text{ mm}$$

$$l_{eff,nc} = 4m + 1.25e = 4 \times 33.65 + 1.25 \times 90 = 247.1 \text{ mm}$$

Row 1 – Effective lengths as first bolt row of a group:

 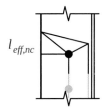

$$l_{eff,cp} = \pi m + p = \pi \times 33.65 + 85 = 190.7 \text{ mm}$$

$$l_{eff,nc} = 2m + 0.625e + 0.5p$$

$$= 2 \times 33.65 + 0.625 \times 90 + 0.5 \times 85 = 166.1 \text{ mm}$$

Row 2 – Individual effective lengths:

 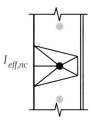

$$l_{eff,cp} = 2\pi m = 2\pi \times 33.65 = 211.4 \text{ mm}$$

$$l_{eff,nc} = 4m + 1.25e = 4 \times 33.65 + 1.25 \times 90 = 247.1 \text{ mm}$$

Row 2 – Effective lengths as first bolt row of a group:

$$l_{eff,cp} = \pi m + p = \pi \times 33.65 + 80 = 185.7 \text{ mm}$$
$$l_{eff,nc} = 2m + 0.625e + 0.5p$$
$$= 2 \times 33.65 + 0.625 \times 90 + 0.5 \times 80 = 163.5 \text{ mm}$$

Row 2 – Effective lengths as last bolt row of a group:

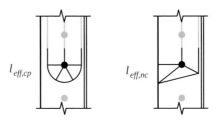

$$l_{eff,cp} = \pi m + p = \pi \times 33.65 + 85 = 190.7 \text{ mm}$$
$$l_{eff,nc} = 2m + 0.625e + 0.5p$$
$$= 2 \times 33.65 + 0.625 \times 90 + 0.5 \times 85 = 166.1 \text{ mm}$$

Row 2 – Effective lengths as internal bolt row of a group:

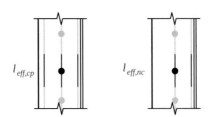

$$l_{eff,cp} = 2p = 2 \times \frac{85 + 80}{2} = 165.0 \text{ mm}$$

$$l_{eff,nc} = p = \frac{85 + 80}{2} = 82.5 \text{ mm}$$

Row 3 – Individual effective lengths:

$$l_{eff,cp} = 2\pi m = 2\pi \times 33.65 = 211.4 \text{ mm}$$

$$l_{eff,nc} = 4m + 1.25e = 4 \times 33.65 + 1.25 \times 90 = 247.1 \text{ mm}$$

Row 3 – Effective lengths as last bolt row of a group:

$$l_{eff,cp} = \pi m + p = \pi \times 33.65 + 80 = 185.7 \text{ mm}$$

$$l_{eff,nc} = 2m + 0.625e + 0.5p$$

$$= 2 \times 33.65 + 0.625 \times 90 + 0.5 \times 80 = 163.55 \text{ mm}$$

The resistances are then determined on the basis of the knowledge of the values of the effective lengths.

Individual resistances of bolt rows 1, 2 and 3:

Mode 1, see Eq. (3.26)

$$F_{T,1,Rd} = \frac{4 \times 3382}{33.7} = 402 \text{ kN}$$

where: $$m_{pl,1,Rd} = \frac{0.25 l_{eff} t_{fc}^2 f_{y,c}}{\gamma_{M0}}$$

$$= \frac{0.25 \times 211.4 \times 16.5^2 \times 235}{1.0 \times 1000} = 3382 \text{ kNmm}$$

Mode 2, see Eq. (3.28)

$$F_{T,2,Rd} = \frac{2 \times 3952 + 42.1 \times 2 \times 254.2}{33.7 + 42.1} = 387 \text{ kN}$$

where: $m_{pl,2,Rd} = \dfrac{0.25 l_{eff,nc} t_{fc}^2 f_{y,c}}{\gamma_{M0}}$

$$= \frac{0.25 \times 247.1 \times 16.5^2 \times 235}{1.0 \times 1000} = 3952 \text{ kNmm}$$

$$F_{t,Rd} = \frac{0.9 \times 1000 \times 353}{1.25 \times 1000} = 254.2 \text{ kN}$$

$$n = \min(e_{min}; 1.25m = 1.25 \times 33.7 = 42.1)$$

$$= 42.1 \text{ mm}$$

Mode 3, see Eq. (3.27)

$$F_{T,3,Rd} = 2 \times 254.2 = 508 \text{ kN}$$

Resistance of combined bolt rows 1 and 2:
In this example, the non-circular yield pattern are decisive.

Mode 1, see Eq. (3.26)

$$F_{T,1,Rd} = \frac{4 \times 5312}{33.7} = 631 \text{ kN}$$

where: $m_{pl,1,Rd} = \dfrac{0.25(166.1 + 166.1) \times 16.5^2 \times 235}{1.0 \times 1000}$

$$= 5312 \text{ kNmm}$$

Mode 2, see Eq. (3.28)

$$F_{T,2,Rd} = \frac{2 \times 5312 + 42.1 \times 4 \times 254.2}{33.7 + 42.1} = 705 \text{ kN}$$

where: $m_{pl,2,Rd} = \dfrac{0.25(166.1 + 166.1) \times 16.5^2 \times 235}{1.0 \times 1000}$

$$= 5312 \text{ kNmm}$$

Mode 3, see Eq. (3.27)

$$F_{T,3,Rd} = 4 \times 254.2 = 1017 \text{ kN}$$

Resistance of combined bolt rows 1, 2 and 3:

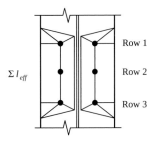

Mode 1, see Eq. (3.26)

$$F_{T,1,Rd} = \frac{4 \times 6591}{33.7} = 784 \text{ kN}$$

where: $m_{pl,1,Rd} = \dfrac{0.25(166.1 + 82.5 + 163.5) \times 16.5^2 \times 235}{1.0 \times 1000}$

$$= 6591 \text{ kNmm}$$

Mode 2, see Eq. (3.28)

$$F_{T,2,Rd} = \frac{2 \times 6591 + 42.1 \times 6 \times 254.2}{33.7 + 42.1} = 1021 \text{ kN}$$

where: $m_{pl,2,Rd} = \dfrac{0.25(166.1 + 82.5 + 163.5) \times 16.5^2 \times 235}{1.0 \times 1000}$

$$= 6591 \text{ kNmm}$$

Mode 3, see Eq. (3.27)

$$F_{T,3,Rd} = 6 \times 254.2 = 1525\,\text{kN}$$

Resistance of combined bolt rows 2 and 3:

Mode 1, see Eq. (3.26)

$$F_{T,1,Rd} = \frac{4 \times 5232}{33.7} = 622\,\text{kN}$$

$$\text{where:} \quad m_{pl,1,Rd} = \frac{0.25(163.5 + 163.5) \times 16.5^2 \times 235}{1.0 \times 1000}$$

$$= 5232\,\text{kNmm}$$

Mode 2, see Eq. (3.28)

$$F_{T,2,Rd} = \frac{2 \times 5232 + 42.1 \times 4 \times 254.2}{33.7 + 42.1} = 703\,\text{kN}$$

$$\text{where:} \quad m_{pl,2,Rd} = \frac{0.25(163.5 + 163.5) + 16.5^2 \times 235}{1.0 \times 1000}$$

$$= 5232\,\text{kNmm}$$

Mode 3, see Eq. (3.27)

$$F_{T,3,Rd} = 4 \times 254.2 = 1017\,\text{kN}$$

e)   Component 5 – End plate in bending

Reference is again made to the equivalent T-stub model. Accordingly, the following geometrical parameters need to be determined:

$e_x = 50 \text{ mm} \quad w = 120 \text{ mm} \quad e = 60 \text{ mm}$

$m_x = u_2 - 0.8a_f\sqrt{2} - e_x = 85 - 0.8 \times 8\sqrt{2} - 50 = 25.9 \text{ mm}$

$m = \dfrac{w}{2} - \dfrac{t_{wb}}{2} - 0.8a_{fb}\sqrt{2} = 60 - 5.1 - 0.8 \times 5\sqrt{2} = 49.3 \text{ mm}$

$m_2 = e_x + e_{1\text{-}2} - u_2 - t_{fb} - 0.8a_{fb}\sqrt{2} = 50 + 85 - 85 - 16 - 0.8 \times 8\sqrt{2} = 24.9 \text{ mm}$

$n_x = \min(e_x; \ 1.25m_x = 1.25 \times 25.9 = 32.4) = 32.4 \text{ mm}$

As previously done, the reduced effect of bolt row 4 is neglected for the estimation of the resistance in bending. The values of the effective lengths are extracted from Table 6.6 of EN 1993-1-8; only group 2-3 is considered here as the presence of the beam flange does not allow developing yield lines between bolt row 1 and bolt row 2. Bolt row 2 needs the estimation of the α coefficient as this bolt row is close to the beam flange; the two following parameters are needed for the estimation of this parameter (see Figure 6.11 of EN 1993-1-8):

$$\left. \begin{array}{l} \lambda_1 = \dfrac{m}{m+e} = \dfrac{49.2}{49.2+60} = 0.45 \\[3mm] \lambda_2 = \dfrac{m_2}{m+e} = \dfrac{24.9}{49.2+60} = 0.23 \end{array} \right\} \quad \rightarrow \alpha = 7.2$$

Row 1 – Individual effective lengths:

$l_{eff,cp,1} = 2\pi m_x = 2\pi \times 25.9 = 163.1 \text{ mm}$

$l_{eff,cp,2} = \pi m_x + w = \pi \times 25.9 + 120 = 201.5 \text{ mm}$

$l_{eff,cp,3} = \pi m_x + 2e = \pi \times 25.9 + 2 \times 60 = 201.5 \text{ mm}$

$l_{eff,nc,1} = 4m_x + 1.25e_x = 4 \times 25.9 + 1.25 \times 50 = 166.3 \text{ mm}$

$l_{eff,nc,2} = e + 2m_x + 0.625e_x = 60 + 2 \times 25.9 + 0.625 \times 50 = 143.2 \text{ mm}$

$l_{eff,nc,3} = 0.5b_p = 0.5 \times 240 = 120 \text{ mm}$

$l_{eff,nc,4} = 0.5w + 2m_x + 0.625e_x = 0.5 \times 120 + 2 \times 25.95 + 0.625 \times 50 = 143.2 \text{ mm}$

Row 2 – individual effective lengths:

$$l_{eff,cp} = 2\pi m = 2\pi \times 49.2 = 309.4 \text{ mm}$$

$$l_{eff,nc} = \alpha m = 7.2 \times 49.2 = 355.0 \text{ mm}$$

Row 2 – effective lengths as first row of a group:

$$l_{eff,cp} = \pi m + p = \pi \times 49.2 + 80 = 234.7 \text{ mm}$$

$$l_{eff,nc} = 0.5p + \alpha m - (2m + 0.625e)$$
$$= 0.5 \times 80 + 7.2 \times 49.2 - (2 \times 49.2 + 0.625 \times 60) = 259.0 \text{ mm}$$

Row 3 – individual effective lengths:

$$l_{eff,cp} = 2\pi m = 2\pi \times 49.2 = 309.4 \text{ mm}$$

$$l_{eff,nc} = 4m + 1.25e = 4 \times 49.2 + 1.25 \times 60 = 272.0 \text{ mm}$$

Row 3 – effective lengths as last row of a group:

$$l_{eff,cp} = \pi m + p = \pi \times 49.2 + 80 = 234.7 \text{ mm}$$
$$l_{eff,nc} = 2m + 0.625e + 0.5p = 2 \times 49.2 + 0.625 \times 60 + 0.5 \times 80 = 176.0 \text{ mm}$$

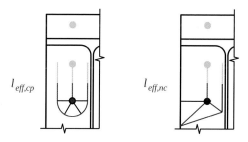

The corresponding resistances are then estimated.

Individual resistances of row 1:

Mode 1, see Eq. (3.26)

$$F_{T,1,Rd} = \frac{4 \times 1586}{25.9} = 245 \text{ kN}$$

where: $m_{pl,1,Rd} = \dfrac{0.25 \times 120 \times 15^2 \times 235}{1.0 \times 1000} = 1586 \text{ kNmm}$

Mode 2, see Eq. (3.28)

$$F_{T,2,Rd} = \frac{2 \times 1586 + 32.4 \times 2 \times 254.2}{25.9 + 32.4} = 337 \text{ kN}$$

where: $m_{pl,2,Rd} = \dfrac{0.25 \times 120 \times 15^2 \times 235}{1.0 \times 1000} = 1586 \text{ kNmm}$

Mode 3, see Eq. (3.27)

$$F_{T,3,Rd} = 2 \times 254.2 = 508 \text{ kN}$$

Individual resistances of row 2:

Mode 1, see Eq. (3.26)

$$F_{T,1,Rd} = \frac{4 \times 4090}{49.2} = 332 \text{ kN}$$

where: $m_{pl,1,Rd} = \dfrac{0.25 \times 309.4 \times 15^2 \times 235}{1.0 \times 1000} = 4090 \text{ kNmm}$

Mode 2, see Eq. (3.28)

$$F_{T,2,Rd} = \frac{2 \times 4693 + 60 \times 2 \times 254.2}{49.2 + 60} = 365 \text{ kN}$$

where: $m_{pl,2,Rd} = \dfrac{0.25 \times 355.0 \times 15^2 \times 235}{1.0 \times 1000} = 4693 \text{ kNmm}$

Mode 3, see Eq. (3.27)

$$F_{T,3,Rd} = 2 \times 254.2 = 508 \text{ kN}$$

Individual resistances of row 3:

Mode 1, see Eq. (3.26)

$$F_{T,1,Rd} = \frac{4 \times 3595}{49.2} = 292.04 \text{ kN}$$

where: $m_{pl,1,Rd} = \dfrac{0.25 \times 272.0 \times 15^2 \times 235}{1.0 \times 1000} = 3595 \text{ kNmm}$

Mode 2, see Eq. (3.28)

$$F_{T,2,Rd} = \frac{2 \times 3595 + 60 \times 2 \times 254.2}{49.2 + 60} = 345 \text{ kN}$$

$$\text{where:} \quad m_{pl,2,Rd} = \frac{0.25 \times 272.0 \times 15^2 \times 235}{1.0 \times 1000} = 3595 \text{ kNmm}$$

Mode 3, see Eq. (3.27)

$$F_{T,3,Rd} = 2 \times 254.2 = 508 \text{ kN}$$

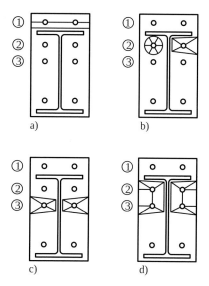

a)

b)

c)

d)

Combined bolt rows 2 and 3:

Mode 1, see Eq. (3.26)

$$F_{T,1,Rd} = \frac{4 \times 5750}{49.2} = 467 \text{ kN}$$

$$\text{where:} \quad m_{pl,1,Rd} = \frac{0.25(259.0 + 176.0) \times 15^2 \times 235}{1.0 \times 1000} = 5750 \text{ kNmm}$$

Mode 2, see Eq. (3.28)

$$F_{T,2,Rd} = \frac{2 \times 5750 + 60 \times 4 \times 254.2}{49.2 + 60} = 664 \text{ kN}$$

$$\text{where:} \quad m_{pl,2,Rd} = \frac{0.25(259.0+176.0)15^2 \times 235}{1.0 \times 1000} = 5750 \text{ kNmm}$$

Mode 3, see Eq. (3.27)

$$F_{T3,Rd} = 4 \times 254.2 = 1017 \text{ kN}$$

f)   Component 6 – Flange and web of the beam in compression

This resistance is estimated as follows, see Eq. (6.20), on the basis of properties extracted from the catalogue of profiles:

$$F_{c,fb,Rd} = \frac{516}{(500-16) \times 10^{-3}} = 1065 \text{ kN}$$

g)   Component 7 – Beam web in tension

a)   b)   c)

This component has to be considered for bolt rows 2 and 3 only. The individual resistances of rows 2 and 3 (case a and b of the figure) and the group resistance (case c of the figure) are given below, see also Eq. (6.22):

Individual resistance of row 2:

$$F_{t,wb,Rd} = \frac{309.4 \times 10.2 \times 235}{1.0 \times 1000} = 742 \text{ kN}$$

Individual resistance of row 3:

$$F_{t,wb,Rd} = \frac{272.0 \times 10.2 \times 235}{1.0 \times 1000} = 652 \text{ kN}$$

Group resistance - rows 2 and 3:

$$F_{t,wb,Rd} = \frac{(259.0 + 176.0) \times 10.2 \times 235}{1.0 \times 1000} = 1043 \text{ kN}$$

g)   Assembling of the components

Force which can be supported by row 1 considered individually

Column flange in bending:	402 kN
Column web in tension:	474 kN
End plate in bending:	245 kN
$F_{1,min} =$	245 kN

Force which can be supported by row 2 considered individually

Column flange in bending:	402 kN
Column web in tension:	474 kN
End plate in bending:	332 kN
Beam web in tension:	742 kN
$F_{2,min} =$	332 kN

Force which can be supported by row 3 considered individually

Column flange in bending:	402 kN
Column web in tension:	474 kN
End plate in bending:	292 kN
Beam web in tension:	652 kN
$F_{3,min} =$	292 kN

275

245 kN

332 kN

292 kN

Figure 6.43 – Forces which can be supported individually by the rows

Account for the group effects (combined rows 1-2 only)

The sum of the individual resistances of rows 1 and 2 is equal to:

$$F_{1,min} + F_{2,min} = 245+332 = 577 \text{ kN}$$

This value is compared to the force which can be supported by bolt rows 1 and 2 acting as a group:

Column flange in bending:	631 kN
Column web in tension:	579 kN
$F_{1+2,min} =$	579 kN

As the group resistance is higher than the sum of the individual resistances, the individual resistances of rows 1 and 2 will be reached (i.e. the maximum force which can be supported by these rows is not affected by the considered group effect).

Account for group effects (combined rows 1-2-3)

The sum of the individual resistances of rows 1-2-3 is equal to:

$$F_{1,min} + F_{2,min} + F_{3,min} = 245 + 332 + 292 = 869 \text{ kN}$$

This value is compared to the force which can be supported by rows 1-2-3 acting as a group:

Column flange in bending:	784 kN
Column web in tension:	653 kN
$F_{1+2+3,min} =$	653 kN

The group resistance is smaller than the sum of the individual resistances. Accordingly, the individual resistances in each row cannot be reached and a reduction of the forces which can be supported as to be taken into account. The force which can be supported by bolt row 3 is estimated as follows:

$$F_{3,min,red} = F_{1+2+3,min} - F_{1,min} - F_{2,min} = 653 - 245 - 332 = 76 \text{ kN}$$

Account for the group effects (combined rows 2-3 only)

The sum of the individual resistances of rows 2-3, taking into account of the reduction estimated in the previous section, is equal to:

$$F_{2,min} + F_{3,min,red} = 332 + 76 = 408 \text{ kN}$$

This value is compared to the force which can be supported by the group formed between rows 2 and 3 estimated as follows:

Column flange in bending:	622 kN
Column web in tension:	573 kN
End plate in bending:	467 kN
Beam web in tension:	1043 kN
$F_{2+3,min} =$	467 kN

As the resistance of the group is higher than the sum of the individual resistances, it is assumed that each individual resistance can be reached.

245 kN

332 kN

76 kN

Figure 6.44 – Maximum forces which can be supported by the bolt rows, taking into account group effects

277

In order to ensure equilibrium of the internal forces in the joint, one has to compare the resistance values of the bolt rows in tension with the other components. The sum of the maximal tensile resistances of rows 1, 2 and 3, taking into account of the group effects, is equal to:

$$F_{1,min} + F_{2,min} + F_{3,min,red} = 244.51 + 332.24 + 76.46 = 653.17 \text{ kN}$$

This value has now to be compared to the following forces:

Column web panel in shear:	549 kN
Column web in compression:	461 kN
Beam flange and beam web in compression:	1065 kN
$F_{glob,min} =$	461 kN

Accordingly, the resistance of the column web in compression limits the maximum forces that can be carried by the bolt rows. The forces in the latter therefore have to be reduced, starting from bolt row 3:

$$F_{1,min} + F_{2,min} = 244.5 + 332 = 577 \text{ kN} > F_{glob,min} = 461 \text{ kN} \Rightarrow F_{3,min,red} = 0 \text{ kN}$$

$$F_{2,min,red} = F_{glob,min} - F_{1,min} = 461 - 245 = 216 \text{ kN}$$

Because the resistance of the bolt row 2 is limited by a "global" component (here: column web in compression), bolt row 3 cannot be considered to contribute to the design moment resistance of the joint. The final summary of the forces which can be supported by the rows is shown in Figure 6.45.

Figure 6.45 – Forces that can be supported by the bolt rows to avoid compression failure of the column web

6.6.3.3 Evaluation of the design moment resistance

The design bending resistance is evaluated as follows:

$$M_{j,Rd} = 245 \times 0.527 + 216 \times 0.442 = 225 \text{ kNm}$$

6.6.3.4 Determination of the rotational stiffness

The first step consists in estimating stiffness coefficients for the different components of the joint, according to Table 6.11 of EN 1993-1-8. These coefficients are given in Table 6.8. The formulae as given in Table 6.11 of EN 1993-1-8 are not repeated in the table.

Table 6.8 – Stiffness coefficients

Components	Stiffness coefficients
Column web panel in shear	$k_1 = \dfrac{0.38 \times 4495}{1.0 \times \left(500 - \dfrac{16}{3} + 85 - 50 - \dfrac{85}{2}\right) \times 10^{-3}} = 3.53$
Column web in compression	$k_2 = \dfrac{0.7 \times 274.8 \times 9.5}{243} = 7.52$
Column web in tension	$k_{3,1} = \dfrac{0.7 \times 166.1 \times 9.5}{243} = 4.54$ $k_{3,2} = \dfrac{0.7 \times 82.5 \times 9.5}{243} = 2.26$ $k_{3,3} = \dfrac{0.7 \times 163.5 \times 9.5}{243} = 4.48$ Contribution from row 4 neglected
Column flange in bending	$k_{4,1} = \dfrac{0.9 \times 166.1 \times 16.5^3}{33.7^3} = 17.62$ $k_{4,2} = \dfrac{0.9 \times 82.5 \times 16.5^3}{33.7^3} = 8.75$ $k_{4,3} = \dfrac{0.9 \times 163.5 \times 16.5^3}{33.7^3} = 17.35$
End plate in bending	$k_{5,1} = \dfrac{0.9 \times 120 \times 15^3}{26.0^3} = 20.86$ $k_{5,2} = \dfrac{0.9 \times 234.7 \times 15^3}{49.2^3} = 5.97$ $k_{5,3} = \dfrac{0.9 \times 176.0 \times 15^3}{49.2^3} = 4.48$
Beam web and beam flange in compression	$k_6 = \infty$
Bolts in tension	$k_{10} = \dfrac{1.6 \times 353}{15 + 16.5 + (15+19)/2} = 11.65$

In a second step, the component assembling is realised in agreement with the recommended rules given in section 6.3.3.1 of EN 1993-1-8. The effective stiffness coefficient of each bolt row is obtained as follows:

$$k_{eff,1} = \cfrac{1}{\cfrac{1}{4.54} + \cfrac{1}{17.62} + \cfrac{1}{11.65} + \cfrac{1}{20.86}} = 2.43 \quad \text{and} \quad h_1 = 527 \text{ mm}$$

$$k_{eff,2} = \cfrac{1}{\cfrac{1}{2.26} + \cfrac{1}{8.75} + \cfrac{1}{11.65} + \cfrac{1}{5.97}} = 1.23 \quad \text{and} \quad h_2 = 442 \text{ mm}$$

$$k_{eff,3} = \cfrac{1}{\cfrac{1}{4.48} + \cfrac{1}{17.35} + \cfrac{1}{11.65} + \cfrac{1}{4.48}} = 1.70 \quad \text{and} \quad h_3 = 362 \text{ mm}$$

The equivalent stiffness coefficient related to the tensile part of the joint is determined as follows:

$$z_{eq} = \frac{2.43 \times 527^2 + 1.23 \times 442^2 + 1.70 \times 362^2}{2.43 \times 527 + 1.23 \times 442 + 1.70 \times 362} = 466.4 \text{ mm}$$

$$\Rightarrow k_{eq} = \frac{2.43 \times 527 + 1.23 \times 442 + 1.7 \times 362}{466.44} = 5.2 \text{ mm}$$

The joint stiffness is then computed by means of Eq. (6.27) of EN 1993-1-8:

$$S_{j,ini} = \frac{210000 \times 466.4^2}{\cfrac{1}{5.23} + \cfrac{1}{7.52} \times \cfrac{1}{3.53}} \times 10^{-6} = 75.2 \text{ MNm/rad}$$

6.6.3.5 Computation of the resistance in shear

The shear force is supported by the four bolt rows. In the case of non-preloaded bolts, the resistance in shear will be limited by the resistance of the bolt shanks in shear or by the bearing resistances of the connected plates (i.e. column flange or end plate). In the previous case, the use of the design rules for these failure modes as recommended in Table 3.4 of EN 1993-1-8 was described; through these rules, it is possible to demonstrate that the failure mode to be considered is the mode bolts in shear:

$$F_{v,Rd} = \frac{\alpha_v f_{ub} A_s}{\gamma_{M2}} = \frac{0.5 \times 1000 \times 353}{1.25} \times 10^{-3} = 141 \text{ kN}$$

Eight bolts contributes to the shear resistance. However, as indicated above, four of them (i.e. the upper two bolt rows) support tensile forces to contribute to the bending resistance. Accordingly, their contribution to the shear resistance has to be limited to take into account of these tensile forces; it is recommended in EN 1993-1-8 to multiply the shear resistance of the bolts by a factor 0.4/1.4 = 0.286. This factor is derived from the tension-shear interaction check assuming the applied tensile force is equal to the tensile resistance of the bolt, see section 3.4.3. The resistance of the joint to shear force is then obtained through the following formula:

$$V_{Rd} = 4F_{v,Rd} + 4F_{v,Rd} \times \frac{0.4}{1.4} = 4F_{v,Rd}\left(1 + \frac{0.4}{1.4}\right) = 726 \text{ kN}$$

UK Specific Comment

It should be noted that European sections are not commonly used in the UK, although there is growing interest in the market. Steel grade S235 is not used in the UK, S355 being the current common grade. Combinations of plates of 15 mm thickness with M24 bolts are common. This combination is assumed to provide enough rotation capacity thus satisfying ductility requirements, because ductile failure of the plates, not brittle failure of the bolts, governs the behaviour.

6.7   APPLICATION TO STEEL COLUMN SPLICES

6.7.1 Common splice configurations

Column splices are required every two or three storeys in a regular building so as to comply with the commercial lengths of the profiles and with the transportation requirements. The splices should hold the connected members in line and transfer the internal forces along the column height.

Splices are predominantly subjected to axial forces, and more particularly to compression. According to the position of the splices along the storey height, the amount of bending moment including second order effects may however vary quite significantly. The designer will usually select the location of the splice so as to decrease the value of the bending moments to be locally transferred.

The detailing of the splice should preferably aim at a perfect continuity of the centroidal axes of the upper column stub, the lower column stub and of the splice material. When this is not the case, the bending moment resulting from the axis eccentricities will have to be explicitly considered in the design. In Figure 6.46, few common "perfectly aligned" splice configurations, taken from Moreno *et al* (2011), are illustrated. They can be divided into two categories: (i) bolted cover plate splices, and (ii) end plate splices.

Figure 6.46 – Common splice configurations (Moreno *et al*, 2011)

For the "bolted cover plate" splices, two situations are considered:

– Bearing type splices (Figure 6.47)

The forces are transferred between the column stubs in direct contact or alternatively through a division plate. Direct bearing does not require the machining or end milling of the columns or even a full contact over the whole column area. A normal preparation is usually acceptable, as long as tolerances specified in EN 1090-2 are respected.

– Non-bearing type splices (Figure 6.48)

The forces are transferred between the column stubs through bolts and cover plates. The forces possibly passing through direct contact are ignored; in most cases, a physical gap between the column stubs is often provided.

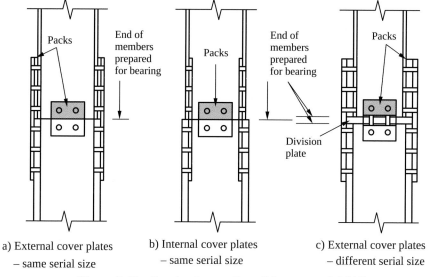

a) External cover plates
– same serial size

b) Internal cover plates
– same serial size

c) External cover plates
– different serial size

Figure 6.47 – Bearing type splices (Moreno *et al*, 2011)

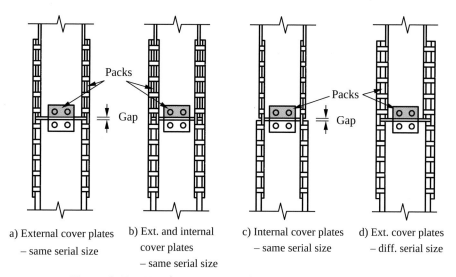

a) External cover plates
– same serial size

b) Ext. and internal
cover plates
– same serial size

c) Internal cover plates
– same serial size

d) Ext. cover plates
– diff. serial size

Figure 6.48 – Non-bearing type splices (Moreno *et al*, 2011)

6.7.2 Design considerations

A splice in a member should be designed to transmit all the moments including second order effects and forces to which the member is subjected at that point. Adequate rotational stiffness should also be provided to the splice so as to validate the design of the column based on member continuity.

Where the members are prepared for full contact in bearing, splice material should be provided to transmit 25% of the maximum compressive force in the column and friction forces between contact surfaces may not be relied upon to hold connected members in place.

Where the members are not prepared for full contact in bearing, splice material should be provided to transmit the internal forces and moments in the member at the spliced section, including the moments due to applied eccentricity, initial imperfections and second-order deformations. The internal forces and moments should be taken as not less than a moment equal to 25% of the moment capacity of the weaker section about both axes and a shear force equal to 2.5% of the normal force capacity of the weaker section in the directions of both axes.

Finally, splices with cover plates used in flexural members should comply with the following requirements:

 – Compression flanges should be treated as compression members;
 – Tension flanges should be treated as tension members;
 – Parts subjected to shear should be designed to transmit the following effects acting together:

  ▪ the shear force at the splice;
  ▪ the moment resulting from the eccentricity of the centroids of the groups of fasteners on each side of the splice, if any;
  ▪ the proportion of moment, deformation or rotations carried by the web or part, irrespective of any shedding of stresses into adjoining parts assumed in the design of the member.

The interested reader will find a large set of interesting information about the detailing and the design of splices in Moreno *et al* (2011).

UK Specific Comment

Standard practice in the UK does not account for this 25% rule. Column splices are usually detailed to SCI (2011) to ensure continuity of stiffness through the splice so that the member design is not invalidated.

6.8   APPLICATION TO COLUMN BASES

6.8.1 Common column basis configurations

Typical column bases are represented in Figure 6.49 (Moreno *et al*, 2011).

Figure 6.49 – Typical column bases (Moreno *et al*, 2011)

They consist of a column, a base plate and holding down bolts (anchors). In Figure 6.49, no stiffeners are added to the steel plate but stiffened base plates may be used where the connection is required to transfer high bending moments, see Figure 6.50. Alternatively the thickness of the base plate could be increased. The column base is usually supported by either a concrete slab or a sub-structure (e.g. a piled foundation) in which the holding down bolts are anchored.

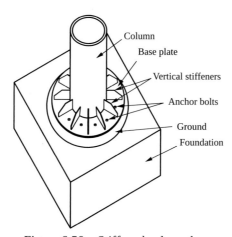

Figure 6.50 – Stiffened column base

285

Different types of anchors may be considered. Some are illustrated in Figure 5.22.

To facilitate a proper erection, the columns are regularly erected on steel levelling packs that are left permanently under the plate. An alternative is to use levelling nuts screwed on the holding bolts before the column is put into position. In all cases, a grout is poured into the space left free between the concrete block and the steel plate.

Other configurations exist, see Figure 6.51 where a so-called embedded configuration is illustrated. Information about this specific solution can be found in Demonceau *et al* (2012).

Figure 6.51 – Embedded column joint configuration

Traditionally column bases are designed as pinned or as rigid joints with a number of anchors chosen accordingly (two for pinned configurations, four for rigid configurations, see Figure 6.52).

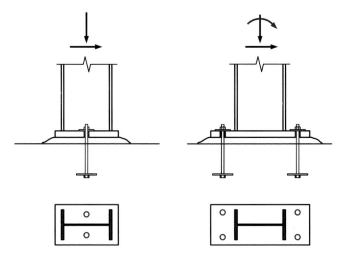

Figure 6.52 – Classical pinned and moment resisting joint configurations

A pinned joint is assumed to resist to axial compression or tension forces and to shear forces, whereas a rigid joint has to additionally transfer bending moments to the foundation.

Compression forces are transmitted to the foundation block by direct bearing and tension forces are transferred through the anchors. Shear forces may be carried out to the concrete block by friction, or directly through the anchors (the rupture of the concrete in bearing and the bearing capacity of the plate should be also checked). For higher levels of shear, designers will often weld a shear stud to the underside of the base plate. Practical considerations about the detailing of column bases may be found in Moreno *et al* (2011).

6.8.2 Design considerations

The following paragraphs are based on a series of papers (Gresnigt *et al*, 2008; Jaspart *et al*, 2008b; Steenhuis *et al*, 2008; Wald *et al*, 2008a; Wald *et al*, 2008b) on which the contents of EN 1993-1-8 are widely based as far as the design of column bases under axial force and bending moment is concerned. For the verification of column base joints under shear forces (topic which is only very briefly addressed in EN 1993-1-8), the reader should refer to Gresnigt *et al* (2008), or section 5.3.4 where the problem has been already addressed.

6.8.2.1 Proportional and non-proportional loading

A characteristic of column bases is that they are loaded by a combination of shear forces, axial forces and bending moments. The loading sequence is also relevant and, from that point of view, two usual cases are here distinguished:

– Non-proportional loading in which the axial force is applied to the base plate connection first, before the moment is increased.
– Proportional loading in which the axial force and the bending moment are applied simultaneously to the joints, with a constant ratio between the bending moment and the axial force, the eccentricity *e*.

In Wald *et al* (2008b), moment-rotation curves obtained for a typical column base with proportional and non-proportional loading are compared. It is shown, for instance, that in the case of non-proportional loading, the initial stiffness of the joint is higher than for non-proportional loading. This is due to the presence of the axial force in the column, which keeps the base plate in contact with the concrete for low bending moments.

6.8.2.2 General procedure for the derivation of the design properties of column bases with base plates

In section 6.3.3, the general procedure to derive the design resistance of a joint subjected to axial forces and bending moments has been described.

It may be applied to column bases, except that, in this case, the position of the compression force is not so obvious to estimate.

Let us first consider a symmetric column base plate joint subjected to pure compression, see Figure 6.53. According to the T-stub component model, the stresses between the plate and the concrete block are assumed to be uniformly distributed on a fictitious rigid equivalent plate resulting from the sum (without overlapping) of the equivalent T-stub rigid equivalent area associated, respectively, to the two column flanges and to the column web, see Figure 6.53.

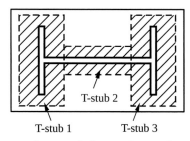

Figure 6.53 – Equivalent rigid plate under axial compression force

The axial resistance N_{Rd} of the column base is simply evaluated as equal to $F_{c,Rd}$ obtained by multiplying the area of the equivalent plate by the design strength of the concrete block in compression (see Figure 6.54).

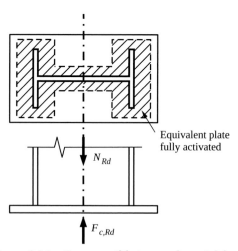

Figure 6.54 – Force equilibrium under axial force

In a second step, a constant axial force N_{Ed} is applied and a bending moment is then progressively increased until the maximum resistance M_{Rd} of the column base joint is reached (non-proportional loading). A specific situation is illustrated in Figure 6.55 where (i) only a part of the rigid equivalent plate, effectively the active part, is subjected to compression $(F_{c,Ed} < F_{c,Rd})$ while (ii) in the tension zone, anchors are in tension and it is assumed that they reach their design resistance, $F_{t,Rd}$.

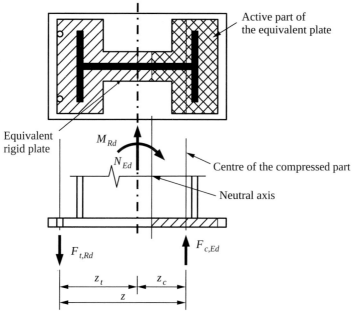

Figure 6.55 – Force equilibrium under axial force and bending moment

By expressing the axial and rotational equilibrium of the forces acting on the column base joints (assembly procedure), the location of the compression force, z_c, and the value of the compression force, $F_{c,Ed}$, will be obtained as well as the resistance of the joint.

The same principles could be applied to the evaluation of the rotational stiffness and axial stiffness of the column base joints.

6.8.2.3 Simplified procedure for the derivation of the design properties of column bases with base plates

The application of the general procedure described above may become rather complex, and hence some authors propose simplified procedures.

In Jaspart and Vandegans (1998), an easy-to-apply calculation procedure is suggested for the evaluation of the design resistance of column base joints. In EN 1993-1-8, another procedure is presented, which is extended to the determination of the joint stiffnesses. The EN 1993-1-8 approach, which covers proportional loading situations, is described below.

It is here suggested to consider that compression is transferred from the column to the concrete block under the column flanges only (Steenhuis, 1998). The effective area under the column web is neglected, as shown in Figure 6.56. The compression forces are assumed to act at the centre of the flanges in compression, also in the cases of the limited outstand of the base, see Figures 6.56c-d. The tensile force is located at the anchor bolts or midway between the bolts when there are two rows of tension bolts, as shown in Figure 6.56a.

The equilibrium of forces may be calculated by referring to Figure 6.56 where two situations are considered as examples: base plate with four anchor bolts (Figure 6.56a), and with two anchor bolts inside the column (Figure 6.56c). $F_{t,l,Rd}$ and $F_{c,r,Rd}$ represent respectively the maximum force that the left anchor bolt row(s) may carry (equal to the weaker components in the tension zone) and the maximum force that than can be transferred to the concrete block by the right column flange in compression ($F_{c,l,Rd}$ on the left side). From Figures 6.56a and c, the following formulae can be derived which express the way on how to check the resistance of the column base:

$$\frac{M_{Ed}}{z} + \frac{N_{Ed}z_{c,r}}{z} \leq F_{t,l,Rd} \tag{6.64}$$

and

$$\frac{M_{Ed}}{z} + \frac{N_{Ed}z_{t,l}}{z} \leq F_{c,r,Rd} \tag{6.65}$$

that can be re-written as:

$$M_{Rd} = \min\left(\frac{F_{t,l,Rd}z}{\frac{z_{c,r}}{e}+1} \quad ; \quad \frac{-F_{c,r,Rd}z}{\frac{z_{t,l}}{e}-1} \right) \tag{6.66}$$

This expression is valid as long as $F_{t,l,Ed}$ and $F_{c,r,Ed}$ are greater than or equal to zero, what may be expressed as follows ($M_{Ed} > 0$ is clockwise and $N_{Ed} > 0$ is tension):

$$N_{Ed} > 0 \text{ and } e = M_{Ed} / N_{Ed} > z_{t,l} \qquad (6.67)$$

$$N_{Ed} \leq 0 \text{ and } e = M_{Ed} / N_{Ed} \geq - z_{c,r} \qquad (6.68)$$

a) two rows of the anchor bolts in tension

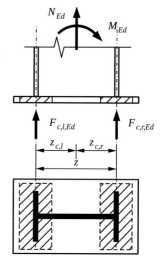

b) no tension in anchor bolts

c) one row of the anchor bolts in tension and limited plate outstand

d) no tension in anchor bolts and limited plate outstand

Figure 6.56 – Equilibrium of forces on the base plate (with the effective area under the flanges only)

A similar reasoning may be applied to the situations illustrated in Figure 6.56b and d where there is no tension force in the anchor bolts but both parts of the joint are under compression. In this case, the equation may be rewritten as:

$$M_{Rd} = \min \left| \frac{-F_{c,l,Rd} z}{\frac{z_{c,r}}{e} + 1} \quad ; \quad \frac{-F_{c,r,Rd} z}{\frac{z_{c,l}}{e} - 1} \right| \qquad (6.69)$$

with the limits:

$$N_{Ed} \leq 0 \text{ and } e = M_{Ed} / N_{Ed} \geq z_{c,l} \qquad (6.70)$$

$$N_{Ed} \leq 0 \text{ and } e = M_{Ed} / N_{Ed} \leq -z_{c,r} \qquad (6.71)$$

Two other cases should be taken into account:

– Left side of the column base in compression and right side in tension;
– Both sides of the column bases are in tension.

The derivation of the related design moment resistances and application fields is immediate. EN 1993-1-8 provides formulae for all cases.

Besides that, similar assumptions than those expressed before may be also considered to evaluate the bending stiffness of a column base plate joint.

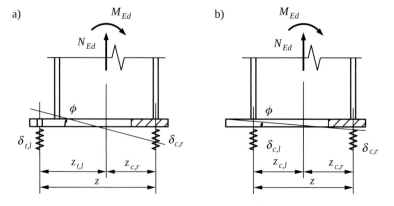

Figure 6.57 – Mechanical stiffness model for a column base plate joint

The elastic deformation of the springs representing the constitutive components may be expressed as follows for the specific situation represented in the left part of Figure 6.57, as an example (Steenhuis, 1998):

$$\delta_{t,l} = \frac{\dfrac{M_{Ed}}{z} + \dfrac{N_{Ed}z_{c,r}}{z}}{Ek_{t,l}} = \frac{M_{Ed} + N_{Ed}z_{c,r}}{Ezk_{t,l}} \tag{6.72}$$

$$\delta_{c,r} = \frac{\dfrac{M_{Ed}}{z} - \dfrac{N_{Ed}z_{t,l}}{z}}{Ek_{c,r}} = \frac{M_{Ed} - N_{Ed}z_{t,l}}{Ezk_{c,r}} \tag{6.73}$$

So the column base rotation writes:

$$\phi = \frac{\delta_{t,l} + \delta_{c,r}}{z} = \frac{1}{Ez^2} + \left(\frac{M_{Ed} + N_{Ed}z_{c,r}}{k_{t,l}} + \frac{M_{Ed} - N_{Ed}z_{t,l}}{k_{c,r}} \right) \tag{6.74}$$

The eccentricity e_0, at which the rotation is zero, can be evaluated from Eq. (6.74) as:

$$\phi = \frac{1}{Ez^2} + \left(\frac{N_{Ed}e_0 + N_{Ed}z_{c,r}}{k_{t,l}} + \frac{N_{Ed}e_0 - N_{Ed}z_{t,l}}{k_{c,r}} \right) = 0 \tag{6.75}$$

Hence the eccentricity under zero rotation is:

$$e_0 = \frac{z_{c,r}\,k_{c,r} - z_{t,l}\,k_{t,l}}{k_{c,r} + k_{t,l}} \tag{6.76}$$

The bending stiffness of the column base plate joint depends on the bending moment due to the change of the eccentricity of the axial force:

$$S_{j,ini} = \frac{M_{Ed}}{\phi} \tag{6.77}$$

and therefore:

$$S_{j,ini} = \frac{M_{Sd}}{M_{Sd} + N_{Sd}e_0} \frac{Ez^2}{\sum \frac{1}{k}} = \frac{e}{e+e_0} \frac{Ez^2}{\sum \frac{1}{k}} \tag{6.78}$$

The reader will find in EN 1993-1-8 similar formulae to cover the whole range of application.

Chapter 7

LATTICE GIRDER JOINTS

7.1 GENERAL

EN 1993-1-8 provides design rules for joints between both open and hollow sections. Design rules for joints between open sections are mainly based on recommendations developed under the umbrella of ECCS, whereas design rules for hollow section joints were published originally as design recommendations of CIDECT. However, these two design methods are based on different approaches: the so-called component method, see 1.6.1, recommended by ECCS is used for the design of joints between open sections and is statistically evaluated against test results, whereas design formulae for hollow section joints are based on semi-empirical investigations in which analytical models were fitted with test results.

Although in many cases the use of hollow sections offers clear advantages, particularly from an architecture perspective, engineers often decide to use open sections instead. There are various reasons for this preference. The design of hollow section joints seems to be complex and only few design tools are available, whereas these are widely available for joints of open sections.

As already mentioned in 1.2.6, design rules for joints between tubular hollow sections are based on theoretical mechanical models and they are then fitted through comparisons with results of experimental and numerical investigations. As a consequence, their field of application is often restricted to the domain for which the rules have been validated, and it cannot be assumed that an extrapolation of the design rules outside the given field of application will lead to safe results. But the range of validity given in chapter 7 of EN 1993-1-8 must be strictly followed.

For different failure modes observed in the experimental tests, the design formulae generally give a resistance value for the joint as a whole, for specific loading cases and specific joint configurations. This restricts the field of application and hence the freedom to modify the joint detailing. It has also to be mentioned that usually no information is provided with regard to the stiffness or the ductility of the joints.

This chapter gives a short introduction to the design of hollow section joints. The interested reader will find a more comprehensive description of the background and the design rules in the CIDECT publication of Wardenier *et al* (2010). This book is freely available on the CIDECT website (http://www.cidect.com). Design tools for hollow section joints (Weynand *et al*, 2011), i.e. design tables for some standardised joints and a special edition of the computer program COP (Weynand *et al*, 2014), are also freely available.

In order to facilitate the use of hollow sections, CIDECT has launched an initiative to adapt the current rules for hollow sections to the same design approach as it is used for open sections (Jaspart and Weynand, 2015) and it is expected that the design rules for both open and hollow sections will be presented in a common approach in a future version of EN 1993-1-8.

7.2   SCOPE AND FIELD OF APPLICATION

Chapter 7 of EN 1993-1-8 gives detailed application rules to determine the static design resistances of uniplanar and multiplanar joints in lattice structures composed of circular, square or rectangular hollow sections. The types of joints covered are shown in Table 1.1. Also, uniplanar joints in lattice structures composed of combinations of hollow sections with open sections are covered to a certain extent. Note that flattened end connections and cropped end connections are not covered.

The application rules are valid both for hot finished hollow sections to EN 10210 (CEN, 2006c, 2006d) and for cold formed hollow sections to EN 10219 (CEN, 2006a, 2006b), if the dimensions of the structural hollow sections fulfil the requirements given in chapter 7 of EN 1993-1-8. Even if the denomination of the sections are the same for hot finished hollow sections according to EN 10210 and cold formed hollow sections according to EN 10219, differences can be found with regard to cross section dimensions and the section properties. Due to this fact, a simple substitution of hot finished sections by cold formed sections cannot be done without taking these differences into account. Furthermore, EN 1993-1-8 gives in section 4.14 restricting conditions for RHS regarding welding in cold-formed zones. Especially for larger brace to chord width ratios, these restrictions have to be kept in mind, whereas for hot finished hollow sections no restrictions exist.

As already said in section 7.1, the range of validity given in chapter 7 of EN 1993-1-8 must be strictly followed. The application rules may only be used if all the conditions specified in EN 1993-1-8 section 7.1.2 are satisfied:

- The compression elements of the members should satisfy the requirements for Class 1 or Class 2 given in EN 1993-1-1 for the condition of pure compression;
- The angles Θ_i between the chords and the brace members, and between adjacent brace members, should satisfy $\Theta_i \geq 30°$;
- The ends of members that meet at a joint should be prepared in such a way that their cross sectional shape is not modified;
- In gap type joints, the gap g between the brace members should satisfy $g \geq t_1 + t_2$, where t_1 and t_2 are the thicknesses of the brace members. This condition ensures that the clearance is adequate for forming satisfactory welds;
- In overlap type joints, the overlap λ_{ov} should be large enough to ensure that the interconnection of the brace members is sufficient for adequate shear transfer from one brace to the other. In any case, the overlap should be $\lambda_{ov} \geq 25\%$. In case the overlap exceeds a certain limit $\lambda_{ov,lim}$, the connection between the braces and the chord should be checked for shear. Detailed limits are given in EN 1993-1-8 clause 7.1.2(6);
- Where overlapping brace members have different thicknesses and/or different strength grades, the member with the lowest ratio t_i/f_{yi} value should overlap the other member.
- Where overlapping brace members are of different widths, the narrower member should overlap the wider one;
- Application limits are given in Tables 7.1 to Table 7.24 of EN 1993-1-8.

Moments resulting from eccentricities in K, N or KT joints may be neglected in the design of tension chord members and brace members. They may also be neglected in the design of connections if the eccentricities are within the following limits:

$$-0.55 d_0 \le e \le 0.25 d_0 \tag{7.1}$$

$$-0.55 h_0 \le e \le 0.25 h_0 \tag{7.2}$$

where:

e is the eccentricity;

d_0 is the diameter of the chord;

h_0 is the depth of the chord, in the plane of the lattice girder.

For the definition of joint eccentricity e, gap g and overlap λ_{ov} see Figure 1.18. The eccentricity depends directly on the gap or overlap, and can be determined from its geometry:

$$e = \left(\frac{2h_1}{\sin \Theta_1} + \frac{2h_2}{\sin \Theta_2} + g \right) \frac{\sin \Theta_1 \sin \Theta_2}{\sin(\Theta_1 + \Theta_2)} - \frac{h_0}{2} \tag{7.3}$$

where:

h_i is the section depth of member i (i = 0, 1 or 2);

Θ_i is the angle of brace members i (i = 1 or 2).

Note: In an overlap joint, the value g is negative.

Hence, the given limitations concerning minimum gap and minimum overlap may lead to the situation that moments resulting from eccentricities may not be neglected. The designer should have in mind this fact when selecting section sizes of the chord and brace members.

298

7.3   DESIGN MODELS

7.3.1 General

The design resistance of joints between braces and chords in lattice structures are expressed in terms of design axial and/or moment resistances of the brace members, or, in some cases, in terms of design axial resistances of the chords.

The complex geometry of the joints, local influences of the corners of rectangular sections and residual stresses, for instance due to welding, lead to non-uniform stress distributions. Strain hardening and membrane effects are also influencing the local structural behaviour. Due to the complexity of the parameters which influence the resistance of the joints, semi-empirical approaches were used to develop design models.

Simplified analytical models which consider the most relevant parameters were developed. Based on extensive research activities, these models were calibrated with parameters based on results of experimental and numerical investigations. The design formulae are based on statistical evaluations against test results. This approach directly led to design values and not to characteristic values. Consequently, the partial safety factor γ_{M5} of EN 1993-1-8 is recommended as $\gamma_{M5} = 1.0$. The design equations are only valid within the investigated parameter ranges. Those application limits are given in EN 1993-1-8, Tables 7.1, 7.8, 7.20 and 7.23 and more specific limits given in the various design resistance tables (for example: Tables 7.3 and 7.4 give additional limits to Table 7.1).

7.3.2 Failure modes

The design resistance of a joint depends on the resistance of the weakest part of the joint. EN 1993-1-8 considers six different failure modes for joints in lattice structures:

- *Chord face failure*, also referred to as *chord plastification*: failure due to plastification of the chord face or the whole chord cross section;
- *Chord web failure*, also referred to as chord side wall failure: failure due to yielding, crushing or instability (crippling or buckling of the chord side wall or chord web) under the compression brace member;
- *Chord shear failure*: failure due to shear forces in chord side wall;
- *Punching shear failure*: chord failure in the flange wall;
- *Brace failure*: failure due to cracking in the welds or in the brace members with reduced effective width;
- *Local buckling failure*: failure due to instability of the chord or brace members at the joint location. However, this failure is not governing for joints within the scope of application according to Tables 7.1 and 7.8 of EN 1993-1-8.

For each failure mode, EN 1993-1-8 provides appropriate design resistance formulae. To determine the design resistance of a joint, all relevant failure modes must be checked and the corresponding resistances should be determined. The minimum resistance is taken as the design resistance of a joint.

299

7.3.3 Models for CHS chords

The design formulae for CHS chord failure modes are based on the so-called *Ring Model* (Togo, 1967). This model assumes that the brace member stresses (axial stresses) into the chord section are transferred mainly at the saddle of the brace members (see point A in Figure 7.1). This loading is taken into account by a pair of concentrated forces acting vertically to the flange, each $0.5 N_1 \sin\Theta_1$. In the longitudinal direction, it is assumed that the loads are transferred within an effective length L_e (see Figure 7.1).

Figure 7.1 – Ring model for chord plastification under axial brace loading (Togo, 1967)

The design formulae are derived assuming that yielding will occur in the chord section. Based on equilibrium of internal and external work, adequate equations can be derived (Togo, 1967). These basic formulae were modified and calibrated based on experimental or numerical investigations (Wardenier, 2001) to cover the various failure modes for relevant practical configurations and loading situations, see Table 7.2 and Table 7.5 of EN 1993-1-8.

7.3.4 Model for RHS chords

As mentioned in section 7.3.2 different failure modes need to be checked to determine the design resistance of a joint. For joints between braces and the flange of rectangular hollow sections, the following failure modes may be relevant: chord face failure, chord web failure, chord shear failure and punching shear failure. The basic models for the design formulae to check those failure modes are briefly described below. More details can be found in Wardenier (1982, 2001), Wardenier *et al* (1991), Packer *et al* (1992), and (Packer and Henderson (1997).

The basic model to check chord face failure is the yield line model for T, Y and X joints under axial load, which is shown in Figure 7.2.

301

Figure 7.2 – Yield line model for T, Y and X joints

By equating the externally applied virtual work to the internal dissipated energy, the lowest load for flange plastification can be calculated. The design values for the resistances for chord face failure given in EN 1993-1-8, Table 7.11 for T, X and Y joints are directly derived from this model. The design models for in-plane bending and out-of-plane bending are derived from that model as well. As the analytical model for N and K joints result in very complex equations, a semi-empirical approach, also based on a yield line model, has been used to develop and validate the design equations in Table 7.12 of EN 1993-1-8.

The model for chord web failure of T, X and Y joints with RHS chords is based on the theory of linear elastic buckling of an isolated. The width of the isolated member is determined by assuming a load distribution into the web of 1:2.5 through the wall thickness. The buckling length is equal to the height of the web. To compute the buckling strength buckling curve a is used. The appropriate design equation is given in EN 1993-1-8, Table 7.11.

In N and K joints with RHS chords, the joint may fail due to high shear in the chord section, see Figure 7.3. In the design equations of EN 1993-1-8, Table 7.12, the influence of the gap size is considered by a factor in the determination of the shear area A_V.

Figure 7.3 – Analytical model for chord shear failure

7.3.5 Punching shear failure

Punching shear failure is also a chord face failure, where the flange of the chord can fail in shear due to the brace loaded in tension, compression or bending. The appropriate formulae in EN 1993-1-8 consider the normal component of the braces $N_i\sin\Theta_i$ and, if necessary, the unequal stiffness distribution along the connection between the brace and the chord flange by the effective width $b_{e,p}$, see Figure 7.4.

Figure 7.4 – Effective width for punching shear failure

7.3.6 Model for brace failure

Due to the unequal stiffness distribution the brace can fail by fracture close to the weld, local yielding or buckling. Like in the equations for punching shear failure, also for brace failure, the effects of the unequal stiffness distribution are taken into account by an effective width b_{eff} in the appropriate formulae of EN 1993-1-8 (Tables 7.10, 7.11, 7.12 and 7.13), see Figure 7.5.

Figure 7.5 – Effective width for brace failure

7.3.7 *M-N* interaction

The design resistances provided in the aforementioned tables of EN 1993-1-8 are resistances of a joint expressed in terms of design axial resistance or design moment resistance of the brace members. When brace member connections are subjected to combined bending and axial force, the following condition should be satisfied.

For welded joints between CHS members:

$$\frac{N_{i,Ed}}{N_{i,Rd}} + \left(\frac{M_{ip,i,Ed}}{M_{ip,i,Rd}}\right)^2 + \frac{\left|M_{op,i,Ed}\right|}{M_{op,i,Rd}} \leq 1.0 \tag{7.4}$$

For welded joints between CHS or RHS brace members and RHS chord members:

$$\frac{N_{i,Ed}}{N_{i,Rd}} + \frac{M_{ip,i,Ed}}{M_{ip,i,Rd}} + \frac{M_{op,i,Ed}}{M_{op,i,Rd}} \leq 1.0 \tag{7.5}$$

where:

$N_{i,Ed}$ is the design value of the internal axial force in member *i* (i = 0, 1, 2 or 3);

$N_{i,Rd}$ is the design value of the resistance of the joint, expressed in terms of the internal axial force in member *i*;

$M_{ip,i,Ed}$ is the design value of the in-plane internal moment in member *i*;

$M_{ip,i,Rd}$ is the design value of the resistance of the joint, expressed in terms of the in-plane internal moment in member *i*;

$M_{op,i,Ed}$ is the design value of the out-of-plane internal moment in member *i*;

$M_{op,i,Rd}$ is the design value of the resistance of the joint, expressed in terms of the out-of-plane internal moment in member *i*.

Chapter 8

JOINTS UNDER VARIOUS LOADING SITUATIONS

8.1 INTRODUCTION

EN 1993 adopts the component method, see section 1.6.1, as a reference for the design of joints in steel structures. Its use enables a wide range of application as far as the individual response of the constitutive components of the studied joint are known and the so-called "assembly procedure of the components" is available. Current knowledge allows a large number of joint configurations, predominatly subjected to bending, to be covered. This chapter reviews recent developments on the application of the component method to various loading situations, including fire, earthquake, impact and explosion.

Guidelines on how to apply the component method for the evaluation of the initial stiffness and the design moment resistance of steel and composite joints are given in EN 1993-1-8. The aspects of ductility are also addressed. The combination of the components proposed in EN 1993-1-8 allows consideration of a wide range of joint configurations and should be largely sufficient to satisfy the needs of practitioners (welded joints, bolted joints with end-plates or cleats, various joint stiffening including transverse column stiffeners, supplementary web plates, backing plates, column web plates and beam haunches). However, assembly design rules are provided for joints under static loading and mainly subjected to bending moments and shear forces, but also for joints connecting profiles with open sections. The application of the component method to joints subjected to axial forces or combinations of axial forces and bending moments has been additionally addressed in sections 6.3.3, 6.3.4 and 6.4.2.

The extension of the application field of the component method may follow different paths, namely: (i) increase the number of components for which design rules are provided to the user, and (ii) derive knowledge for the characterisation of the component and the assembly of components under other loading conditions. Through (i), the field will be extended to other joint configurations. Weynand *et al* (2015) gathered relevant information for many different components for which design rules have been recently developed. In the same reference, the practical application of the component method to tubular joints is also covered, including component rules, design recommendations, and worked examples.

In the following sections, it is intended to investigate option (ii) further, and present recent work with this aim.

8.2 JOINTS IN FIRE

Simões da Silva *et al* (2011) first proposed to refer to the component method to characterise the behaviour of steel joints at high temperature. In their study, they have demonstrated that the rotational stiffness and the bending resistance of a structural joint may be simply obtained by multiplying the corresponding properties derived at room temperature by ad-hoc reduction factors (respectively $k_{E;\theta}$ for stiffness and $k_{y;\theta}$ for resistance) evaluated according to EN 1993-1-2 (CEN, 2005b). These factors express the decrease of the steel Young modulus E and yield strength f_y at temperature θ. However, this simple approach is limited to cases where all the components are subjected to the same increase of temperature, which is not often reflecting the reality. This is why in Demonceau *et al* (2013) the procedure has been improved as follows: (i) in the tests used as references, temperatures θ_i have been measured in all components i, (ii) the stiffness and resistance properties of components i have been multiplied by ad-hoc reduction factors $k_{E;\theta}$ and $k_{y;\theta}$, and (iii) the the components have been assembled to characterise the global response of the joint, paying attention to the variation of temperature in the joint. The procedure has been validated with test results on composite joints. To allow an easy application in practice, Demonceau *et al* (2009) have proposed replacing the measurement of actual temperatures in the joint during laboratory tests by a thermal analysis with SAFIR (Franssen, 2005). In some cases, analytical expressions to determine the temperature of individual components as a function of the time have been proposed.

8.3   JOINTS UNDER CYCLIC LOADING

In EN 1998-1-1 (CEN, 2004c), and in particular in chapters 6 and 7 dealing respectively with the seismic design of steel and composite structures, it is clearly stated that the use of partial strength joints is permitted but the number of requirements to be respected for this joint typology are such that it is almost compulsory to perform physical tests to check when these requirements are fulfilled.

Accordingly, the use of partial strength joints in structures prone to seismic actions is very limited and, as a consequence, the practitioners have to design full strength joints taking into account the possible overstrength effects, which leads to expensive joint solutions and so limits the competiveness of steel structures compared to other structural solutions. An example of such optimised full strength joint solution is presented in Figure 8.1 (Hoang *et al*, 2014); this solution was developed in the framework of a recent RFCS project entitled HSS-SERF (High Strength Steel in SEismic Resistant building Frames) (Dubina, 2015).

307

The component method could be a solution to overcome this "full-strength" obligation. Indeed, the component method has the potential to predict the response of joints under cyclic loading but for that, it is required to know the behaviour of each component under such loading conditions. In particular, it is necessary to know the post-yielding behaviour of the components accounting for the strain-hardening effects, their ultimate resistance, their deformation capacity but also the degradation of their strength and stiffness under the applied cycles associated to phenomena of oligo-cyclic fatigue. In addition, it is needed to know the unloading behaviour of each component as such unloading may occur during the cyclic loading imposed by the seismic action.

Investigations have been recently conducted at the University of Coimbra (Nogueiro *et al*, 2009) with very promising results. The proposed model consists in a numerical implementation of a hysteretic model able to simulate a generic steel joint behaviour. The use of a numerical implementation is required as the evolution of the loads in each component at each step of the applied loading has to be known in order to be able to detect the strength/stiffness degradation and the moment at which the maximum deformation capacity of a component is reached. The proposed model in Nogueiro *et al* (2009) is still under development/improvement nowadays, in particular through contributions to the European RFCS project EQUALJOINTS (European pre-QUALified steel JOINTS).

Elements		Steel materials
1	I-steel beam	Mild steel
2a, 2b	Top and bottom hammer-head	Extracted from the beam profiles
3	Partially-encased wide-flange	High strength steel may be used
4	End plate	Mill steel
5	Bolts	High strength bolts (8.8 or 10.9)
6	Lateral plates	Same grade with the column profiles

Figure 8.1 – Full strength optimised beam-to-column joint solution

8.4 JOINTS UNDER EXCEPTIONAL EVENTS

A structure should be designed to behave properly under service loads (at SLS) and to resist design factored loads (at ULS). The type and the intensity of the loads to be considered in the design process may depend on different factors such as: the intended use of the structure (e.g. type of variable loads), the location (e.g. wind action, level of seismic risk), and even the risk of accidental loading (e.g. explosion, impact, flood). In practice, these individual loads are combined so as to finally derive the relevant load combination cases. In this process, the risk of an exceptional (and therefore totally unexpected) event leading to other accidental loads than those already taken into consideration in the design process in itself is not at all covered. This is a quite critical situation in which the structural integrity should be ensured, i.e. the global structure should remain globally stable even if one part of it is destroyed by the exceptional event (e.g. explosion, impact, fire as a consequence of an earthquake). In conclusion, structural integrity is required when the structure is subjected to exceptional actions not explicitly considered in the definition of the design loads and load combination cases.

Under such exceptional actions, the structural elements, and the joints, in particular, are generally subjected to loadings not initially foreseen through the ULS design. For instance, if the exceptional event "loss of a column" is considered, joints that are initially designed to transfer shear forces and hogging bending moments will experience high tying forces after the loss of a column, as a result of the development of membrane forces in the beams located just above the damaged or destroyed column. Moreover, a reversal of moments occurs in the joints located just above the damaged column. Finally, the joints could be subjected to some dynamic effects if the column loss is for instance induced by an impact or an explosion.

In section 6.3.3, it has been shown how the $M - N$ plastic resistant curve of a joint can be predicted through the use of the component method. Of course, the methodology presented there can help predicting the behaviour of the joints when subjected to exceptional events. However, it has to be pointed out that, when considering the behaviour of structures subjected to such event, the main objective is to ensure that the building will remain globally stable and so it can be accepted to go a step further in the resistance of the structural elements in comparison to what is imposed as limits for ULS. Accordingly, in addition to the prediction of the plastic resistance of the joints under $M - N$, it is also important to be able to predict the ultimate resistant curve, i.e. to be able to predict the $M - N$ combinations under which the joints fail. Jaspart and Demonceau (2008a) explain how the $M - N$ plastic and ultimate resistance curves can be predicted for composite joints and how the proposed model has been validated with experimental results.

In parallel with the prediction of the resistance curves, another key issue to be considered when predicting/estimating the robustness of a structure is the prediction of the deformation capacity or the ductility of a joint. Research efforts are still required in this field. Finally, another aspect to be dealt with when considering the behaviour of joints under exceptional events is the possible dynamic effects which can be induced by these events, dynamic effects which can be associated to strain rate effects in some joint components. This aspect is presently under investigation at the University of Liège in the framework of a RFCS European project entitled ROBUSTIMPACT.

UK Specific Comment

The UK Building Regulations require that joints are designed for tying forces in order to satisfy structural integrity requirements. Guidance on the design values of tying forces is given in BS EN 1991-1-7 Annex A and its UK NA (BSI, 2014).

Chapter 9

DESIGN STRATEGIES

9.1 INTRODUCTION

This chapter describes the various approaches that can be used to design steel frames with due attention being paid to the behaviour of the joints. In practice, this design activity is normally performed by one or two parties, according to one of the following ways:

- An engineering office (in short *engineer*) and a steel fabricator (in short *fabricator*), referred as *Case A*;
- An engineering office (*engineer*) alone, referred as *Case B1*;
- A steel fabricator (*fabricator*) alone, referred as *Case B2*.

At the end of this design phase, fabrication by the steel fabricator takes place. The share of responsibilities for design and fabrication respectively is given in Table 9.1 for these three cases.

311

Table 9.1 – Parties and their roles in the design/fabrication process of a steel structure

Role	Case A	Case B1	Case B2
Design of members	Engineer	Engineer	Fabricator
Design of joints	Fabricator	Engineer	Fabricator
Fabrication	Fabricator	Fabricator	Fabricator

The design process is ideally aimed at ensuring that a given structure fulfils architectural requirements, on one hand, and is safe, serviceable and durable for a minimum of overall cost, on the other hand. The parties involved in the design and fabrication activities also care about optimising their respective profits.

In *Case A*, the engineer designs the members, while the steel fabricator designs the joints. It is up to the engineer to specify the mechanical requirements to be fulfilled by the joints. The fabricator has then to design the joints accordingly, keeping in mind the manufacturing aspects also. Due to the disparity in the respective involvements of both parties, the constructional solution adopted by the fabricator for the joints may be sub-optimal; indeed it is dependent on the beam and column sizing that is made previously by the engineer. For example, if the beams and columns are of minimum size, the joints may then need stiffeners in order to achieve safety and serviceability requirements. If larger members are specified, then joints may prove to be less complex and result in a better economy of the structure as a whole (Figure 9.1).

In *Case B1*, the engineer designs both members and joints. He is thus able to account for mechanical joint properties when designing members. He can search for global cost optimisation too. It may happen however that the engineer has only a limited knowledge of the manufacturing requisites (machinery used, available materials, bolt grades and spacing, accessibility for welding, etc.); then this approach may contribute to some increase in the fabrication costs.

a) Bolted end-plate with haunch   b) Bolted flush end-plate

Figure 9.1 – Two solutions: different economy

*Case B*2 is ideal with regard to global economy. Indeed, the design of both members and joints are in the hands of the fabricator who will be well aware of all the manufacturing aspects.

Before commenting on these various approaches, it is necessary to introduce some definitions. A joint is classified as *simple, semi-continuous* or *continuous.* This wording is general. It is concerned with resistance, with stiffness or with both. In relation to methods of global frame analysis, this wording leads to more commonly used terms:

1. In an elastic global frame analysis, only the stiffness of the joints is relevant. Then, a simple joint is a *pinned* joint, a continuous joint is a *rigid* joint , and a semi-continuous joint is a *semi-rigid* joint;

2. In a rigid-plastic analysis, only the resistance of the joints is relevant. Then, a simple joint is a *pinned* joint, a continuous joint is a *full-strength* joint while a semi-continuous joint is *a partial strength* joint.

The various cases described above are considered below. For the sake of simplicity, it is assumed that global frame analysis is conducted based on an elastic method of analysis. This assumption is not a restriction, though; similar conclusions could be drawn if another type of analysis is performed.

For the design of steel frames, the designer can adopt one of the following design approaches:

Traditional design approach: the joints are assumed to be either simple or continuous. The members are designed first, and then the joints. Such an approach may be used in any Case A, B1 or B2; it is common practice in almost all European countries.

Integrated design approach: both member and joint properties are accounted for when starting the global frame analysis, which can therefore reflect the interaction between members and joints. This approach is normally used in Cases B1 and B2, and possibly in Case A.

313

UK Specific Comment

In the UK the "fabricators" are known as "Steelwork contractors" because they do more than just fabricate.

9.2   TRADITIONAL DESIGN APPROACH

In the traditional design approach, any joint is assumed to be either simple or continuous. A simple joint is capable of transmitting the internal forces determined from the global frame analysis but does not develop a significant moment resistance which might affect the beam and/or column behaviour. A continuous joint permits only a limited rotation between the connected members as long as the applied bending moment does not exceed the bending resistance of the joint.

The assumption of simple and/or continuous joints allows sharing of design activities into two more or less independent tasks, with limited data flow in between. The traditional design approach of any steel frame consists of eight steps, see Figure 9.2.

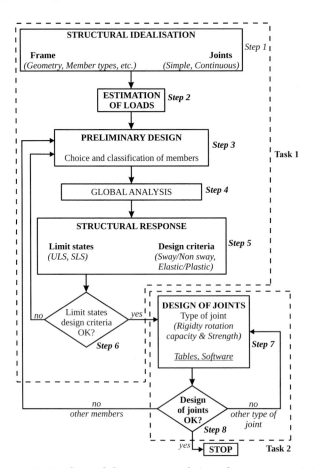

Figure 9.2 – Traditional design approach (simple/continuous joints)

Step 1: The structural idealisation is a conversion of the real properties of the frame into the properties required for frame analysis. Beams and columns are normally modelled as bars. Depending on the type of frame analysis which will be applied, properties need to be assigned to these bars. For example, if an elastic analysis is used, only the stiffness properties of the members are relevant. The joints are pinned or rigid and are modelled accordingly (Figure 9.3).

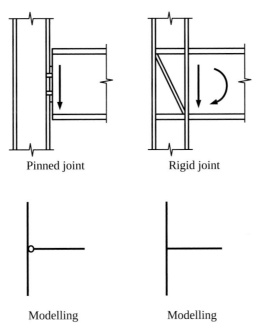

Pinned joint   Rigid joint

Modelling   Modelling

Figure 9.3 – Modelling of pinned and rigid joints (elastic global analysis)

Step 2: Loads are determined based on relevant standards.

Step 3: The designer normally performs the preliminary design, or pre-design, of beams and columns by taking advantage of his own design experience from previous projects, or simple design rules, for the initial sizing of the members. In the pre-design, assumptions have to be made concerning the stress distributions within the sections (elastic, plastic), and any allowances for plastic redistribution between sections. These assumptions will affect the classes of members that may be used. The validity of this assumption has to be verified later in Step 5.

Step 4:   The input for global frame analysis depends on the type of analysis. In an elastic analysis, the input is the geometry of the frame, the loads and the flexural stiffness of the members. In a rigid-plastic analysis, the input is the geometry of the frame, the loads and the resistance of the members. An elastic-plastic analysis requires both resistance and flexural stiffness of the members. Whatever the type of global frame analysis, the distribution and the magnitude of the internal forces and displacements are the output (however rigid-plastic analysis does not allow for any information regarding displacements).

Step 5:   Limit state verifications consist normally in checking the displacements of the frame and of the members under service loading conditions (SLS), the resistance of the member sections (ULS), as well as the frame and member stability (ULS). The assumptions made regarding the section classes (see Step 3) are also checked.

Step 6:   Chosen member sizes will need to be refined when the limit state verifications fail, or when the utilisation of members is shown to be low (the choice is too conservative). Member sizing is adjusted by choosing larger shapes in the first case, smaller ones in the second case. Normally the designer's experience and know-how form the basis for the decisions made in this respect.

Steps 7/8: The member sizes and the magnitude of the internal forces that are applied to the joints are the starting point for the design of joints. The purpose of any joint design is to find a configuration which allows a safe transmission of the internal forces between the connected members. Additionally, when a simple joint is adopted, it is essential that the detailing ensures that no significant bending moment develops in the joint. For a continuous joint, the designer has to check whether the joint satisfies the assumptions made in Step 1 (for instance, whether the joint stiffness is sufficiently large when an elastic global frame analysis is performed). In addition, the rotation capacity has to be appropriate, for example the detailing of simple joints must ensure they are ductile.

The determination of the mechanical properties of a joint is called *joint characterisation* (see sub-chapter 1.6). To check whether a joint may be classified as simple or continuous, reference is made to this *joint classification* (see sub-chapter 2.4).

Rules for joint characterisation in compliance with EN 1993 are available (see chapters 3 to 6), but design tools can be very helpful during the design process because they greatly facilitate this task. In many cases the designer may select the joints from tables which provide the strength and stiffness properties of the relevant joints, as well as, when necessary, the rotation capacity. Dedicated software may be an alternative to the latter. They require the whole layout of the joint as input and provide strength, stiffness and rotational capacity as output. Computer based design proceeds interactively by trial and error, starting with a simple (lowest cost) solution.

The mechanical properties of a joint shall be consistent with those assumed in the modelling of this joint in the global frame analysis. Either the design of the joint and/or that of the members may need to be adjusted.

9.3   INTEGRATED DESIGN APPROACH

In the integrated design approach, the global analysis is carried out taking into account the presumed real joint response (Figure 9.4). It is therefore different from the traditional design approach in several respects:

- *Structural concept:* in the structural concept phase, the real mechanical behaviour of the joints is modelled;
- *Preliminary design:* in the pre-design phase, joints are selected by the designer based on his experience. Details of the joint are determined: end plate or cleat dimensions, location of bolts, number and diameter of bolts, sizes of column and beam flanges, thickness and depth of column web, etc.;
- *Determination of the mechanical properties:* in Step 4, the structural response of both the selected members and joints is determined. First the joints are characterised (see sub-chapter 1.6) because characterisation may identify non-linear behaviour. It is followed by an idealisation, for instance according to a linear or bi-linear joint response curve (see sub-chapter 2.3), which becomes a part of the input for global frame analysis;

   – *Global frame analysis:* for the purpose of global frame analysis, the joint structural response is represented by a relevant spring in the frame model. This activity is called *modelling* (see sub-chapter 2.2).

Of course an integrated design approach is only practical when both members and joints are designed by a single party, because the mechanical properties of the joints must be accounted for when starting the global frame analysis. As it accounts for any type of behaviour, the integrated design approach is especially applicable to frames with so-called semi-continuous joints. The modelling of semi-continuous joints is presented in sub-chapter 2.2 in relation to the global frame analysis.

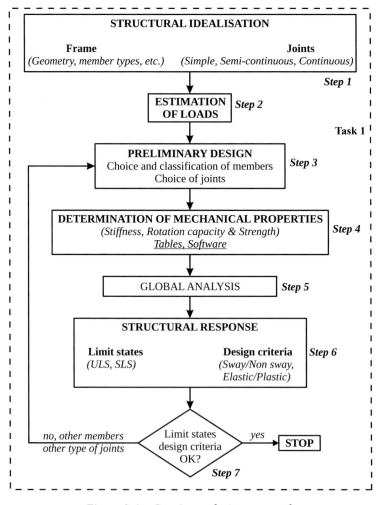

Figure 9.4 – Consistent design approach

9.4   ECONOMIC CONSIDERATIONS

9.4.1 Fabrication and erections costs

a)   Optimal detailing of rigid joints

A very efficient strategy can be summarised as *optimising the joint detailing such that the joint stiffness comes close to the 'rigid' classification boundary, but remains higher.* This is illustrated in Figure 9.5.

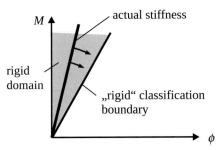

Figure 9.5 – Optimization of rigid joints

The actual stiffness of a joint as well as the classification boundary for rigid joints can be calculated according to EN 1993-1-8. The classification boundary is the minimum stiffness required to model a joint as rigid. If the actual joint stiffness is significantly higher, e.g. due to stiffeners, it should be checked whether it is possible to simplify the joint in order to reduce fabrication costs, while still fulfilling the criterion for rigid joints. This procedure is used in the example described below.

The joints of a typical portal frame (see Figure 9.6) were designed using traditional design practice.

span: 20.0 m, crane girder at 5 m height
roof slope: 2-3%, frame distance: 6.0 m
clearance: 7.0 m, column bases: rigid

Figure 9.6 – Example for the optimization of rigid joints

319

To classify the joints as rigid in an unbraced frame, EN 1993-1-8 requires that:

$$S_{j,ini} \geq 25 \frac{EI_b}{L_b} = 86 \text{ kNm/mrad}$$

According to EN 1993-1-8 the joint characteristics in Figure 9.6 are calculated as follows:

- design moment resistance $M_{j,Rd} = 282$ kNm
- initial stiffness $S_{j,ini} = 145$ kNm/mrad

Hence the joint is classified as rigid. But in order to optimise the joint, the detailing of the joint ican be modified step by step as follows:

1. omit the stiffeners at compression side,
2. in addition, omit the stiffeners at tension side (top of the column),
3. in addition, omit the lowest bolt row in tension.

For all variations it is requested that the joint may still be classified as rigid, i.e. the initial stiffness of the joint should be greater than 86 kNm/mrad. Table 9.2 gives the resistance and stiffness for all variations. It can be seen that the design moment resistance is less than the applied moment and the joint behaves as rigid. The different joint details will have no influence on the design of the member as the joint remains rigid. Therefore, differences in the fabrication costs of the joints gives a direct indication of the economic benefits. The indicative fabrication costs (material and labour) are given for all investigated solutions in Table 9.2 as a percentage of the fabrication costs of the original joint layout. Beside the reduction of fabrication costs, the modified joints will be more ductile. Another possibility to optimize the joint is the use of thinner plates. This can also lead to a more ductile behaviour. Other advantages are: smaller welds, the preparation of hole drilling is easier or hole punching becomes possible.

320

Table 9.2 – Variations in joint detailing and relative savings in fabrication costs

Variations	$M_{j,Rd}$ [kNm]	$S_{j,ini}$ [kNm/mrad]	Stiffness classification	Relative fabrication costs [*)]	Saving
stiffener / IPE 360 / HEA 300	255.0	93	rigid	87 %	13 %

Table 9.2 – Variations in joint detailing and relative savings in fabrication costs
(continuation)

Variations	$M_{j,Rd}$ [kNm]	$S_{j,ini}$ [kNm/mrad]	Stiffness classification	Relative fabrication costs [*)	Saving
no stiffener / IPE 360 / HEA 300	250.6	89	rigid	73 %	27 %
IPE 360 / bolt row ommited / HEA 300	247.8	88	rigid	72 %	28 %

[*) fabrication costs of joints relative to those of the configuration in Figure 9.6

b)   Use of semi-rigid joints

Semi-rigid joints may be used in order to achieve an optimal balance between member sizes and joint details. The ideal assumption that the joints are rigid can often lead to situations where it is not possible to avoid stiffeners. As a consequence the joints have high fabrication costs (e.g. welding). More economical solutions can be found by crossing the rigid classification boundary to semi-rigid joints (see Figure 9.7).

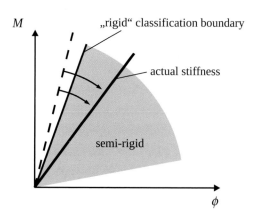

Figure 9.7 – Optimisation with semi-rigid joints

321

The detailing of the joints is then chosen for economy. The semi-rigid joint behaviour has to be taken into account in the frame analysis. This is possible by modelling the joints with rotational springs at the beam ends. Their behaviour influences the global response of the frame, i.e. moment redistributions and displacements. As a consequence the size of some members may increase in comparison with a design with rigid joints. The optimum solution can only be found when a detailed calculation of the costs is carried out. The use of joints without stiffeners may provide more advantages than just reduced costs: it becomes easier to connect secondary beams, service pipes may be installed between the column flanges, easier application of coating and fewer problems with corrosion. The steel construction can appear more aesthetic and lighter, and joints without stiffeners are usually more ductile. With respect to ULS design, this is an important aspect, particularly when considering seismic or robustness requirements.

9.4.2 Savings of material costs

Moment connections are usually required for unbraced frames. For braced frames simple joints are normally more economic, although the moment diagram for the (simply supported) beams leads to beam sizes which are optimised for the mid-span moment, but oversized at their ends. Also, because the joints do not transfer any moment, it is sometimes necessary to install additional temporary bracing during erection. However, in many cases, joints which are assumed to behave as nominally pinned (e.g. flush end plates) actually behave as semi-rigid joints. Because the joints may have some inherent stiffness and may transfer moments, a strategy can be to take advantage of that actual behaviour. This can improve the distribution of moments and allow smaller members (also for the bracing systems) without any change in joint detailing. A further strategy could be to check if small changes to simple joints could have a disproportionate benefit to the frame.

Sometimes it will be possible to improve simple joints without a large increase in fabrication cost (e.g. flush end plates instead of a partial depth end plates). As such joints can transfer significant moments, the bending diagram will be more balanced and the members can be smaller. Clearly, the increase in fabrication costs for the joints and decrease of weight of the structure are in competition and a check of balance is necessary. However recent investigations show that the most economical solutions can be found if the contributions of the stiffness and resistance of the joints are considered, i.e. semi-rigid joints are used.

9.4.3 Summary and conclusions

With respect to joint design basically two different strategies can be identified when targeting minimum cost steel structures:

– Simplification of the joint detailing, i.e. reduction of fabrication costs. Typically this is relevant for unbraced frames when the joints transfer significant moments (traditionally rigid joints);
– Reduction of member dimensions, i.e. reduction of material costs. Typically this is relevant for braced frames with simple joints.

In general both strategies would benefit from the use of semi-rigid joints. Economy studies in various countries have shown possible benefits from the use of semi-rigid joints. More significant savings can be achieved when moment connections are optimised in view of economy. It is interesting to note that all the studies for similar values for the saving due to the use of semi-rigid joints was compared to traditional design solutions. From the different studies it can be concluded that the possible savings due to semi-rigid design can be 20 – 25 % in the case of unbraced frames and 5 – 9 % in the case of braced frames. With the assumption that the cost of the frame is about 10% of the total cost for office buildings, and about 20 % for industrial buildings, the reduction of the total building costs could be estimated as 4 – 5% for unbraced frames. For braced systems savings of 1 – 2% are possible.

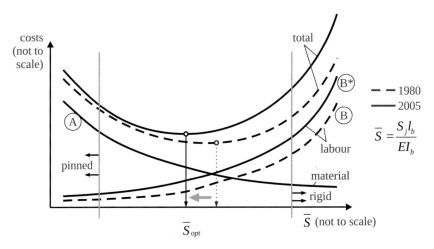

Figure 9.8 – Costs of steel structures depending on the relative joint stiffness

The following general conclusion can be drawn from the investigations presented in this chapter. As shown in Figure 9.8, the costs for material and fabrication (labour) are dependent on the relative stiffness of the joints. While the material costs decrease (curve A), the labour costs increase (curve B) with an increasing joint stiffness. From the sum of these curves a minimum total cost can be found and, from this, an optimum joint stiffness. Following the tendency of the last decades, it is obvious that labour costs are increasing in comparison to the material costs (see curve B*). From Figure 9.8 it becomes clear that as a consequence there is a progressive evolution of the optimum joints stiffness towards more flexible joints. Hence, to find economical solutions for steel structures the use of semi-rigid joints will be of growing interest.

BIBLIOGRAPHIC REFERENCES

Anderson D, Aribert J-M, Bode H, Huber G, Jaspart J-P, Kronenberger H-J, Tschemmernegg F (1999). *Design of composite joints for buildings*, edited by ECCS TC 11, Publ. No. 109, ECCS - European convention for constructional steelwork, Brussels, Belgium.

Astaneh A, Bergsma G, Shen JR (1992). *Behavior and Design of Base Plates for Gravity, Wind and Seismic Loads*. In *National Steel Construction Conference*, edited by AISC. Las Vegas.

BCSA/SCI Connections Group (1992). *Joints in Steel Construction: Volume 2: Practical Applications*, 2nd Ed., in: SCI Publ. P206, The Steel Construction Institute and The British Constructional Steelwork Association.

BCSA/SCI Connections Group (1993). *Joints in Steel Construction: Volume 1: Design Methods*, 2nd Ed., in: SCI Publ. P205, The Steel Construction Institute and The British Constructional Steelwork Association.

BCSA/SCI Connections Group, Hole E, Malik A, Way A (2002). *Joints in Steel Construction: Simple Connections*, SCI Publ. P212, The Steel Construction Institute and The British Constructional Steelwork Association.

Bouwman LP, Gresnigt N, Romeijn A (1989). *Onderzoek naar de bevestiging van stalen voetplaten aan funderingen van beton. (Research into the connection of steel base plates to concrete foundations)*, Report 25.6.89.05/c6, TU-Delft Stevin Laboratory, Delft.

Brettle M (2008). *Steel Building Design: Worked Examples - Hollow Sections*, SCI Publ. P374, The Steel Construction Institute.

Brettle M. (2009). *Steel Building Design: Worked Examples - Open Sections*, SCI Publ. P364, The Steel Construction Institute.

Brown D, Iles D, Brettle M, Malik A, BCSA/SCI Connections Group (2013)/ SCI (2013a). *Joints in Steel Construction: Moment-Resisting Joints to Eurocode 3*, SCI Publ. P398, The Steel Construction Institute and The British Constructional Steelwork Association.

BSI (2000). BS 5950-1:2000 – Structural use of steelwork in building - Part 1: Code of practice for design. Rolled and welded sections, *British Standards Institution,* United Kingdom.

BSI (2007). BS EN 14399-8:2007 – High-strength structural bolting assemblies for preloading. System HV. Hexagon fit bolt and nut assemblies, *British Standards Institution,* United Kingdom.

BSI (2009). BS EN 1993-1-12:2007 – Eurocode 3: Design of steel structures – Part 1.12: Additional rules for the extension of EN 1993 up to steel grades S 700, *British Standards Institution,* United Kingdom.

BSI (2010). BS EN 1993-1-8:2005 – Eurocode 3: Design of steel structures – Part 1.8: Design of joints, *British Standards Institution,* United Kingdom.

BSI (2013). BS EN ISO 898-1:2013 – Mechanical properties of fasteners made of carbon steel and alloy steel. Bolts, screws and studs with specified property classes. Coarse thread and fine pitch thread, *British Standards Institution,* United Kingdom.

BSI (2014). BS EN 1991-1-7:2006+A1:2014 – Eurocode 1. Actions on structures – Part 197: General actions, *British Standards Institution,* United Kingdom.

BSI (2015). BS EN 14399-4:2015 – High-strength structural bolting assemblies for preloading. System HV. Hexagon bolt and nut assemblies, *British Standards Institution,* United Kingdom.

CEB (1994). *Fastenings to Concrete and Masonry Structures - State of the Art Report,* edited by CEB, (Comité Euro-International du Béton), Thomas Telford Services Ltd, London.

CEB (1996). *Design of fastenings in concrete - Design Guide,* edited by CEB, (Comité Euro-International du Béton), Thomas Telford Services Ltd, London.

CEN (2002). EN 1990:2002 – Eurocode - Basis of structural design, *European Committee for Standardization,* Brussels, Belgium.

CEN (2004a). EN 1992-1-1:2004 – Eurocode 2: Design of concrete structures - Part 1-1: General rules and rules for buildings, *European Committee for Standardization,* Brussels, Belgium.

CEN (2004b). EN 1994-1-1:2004 – Eurocode 4: Design of composite steel and concrete structures - Part 1-1: General rules and rules for buildings, *European Committee for Standardization*, Brussels, Belgium.

CEN (2004c). EN 1998-1-1:2004 – Eurocode 8: Design of structures for earthquake resistance - Part 1: General rules, seismic actions and rules for buildings, *European Committee for Standardization*, Brussels, Belgium.

CEN (2004d). EN 10025-1:2004 – Hot rolled products of structural steels - Part 1: General technical delivery conditions, *European Committee for Standardization*, Brussels, Belgium.

CEN (2004e). EN 10164:2004 – Steel products with improved deformation properties perpendicular to the surface of the product - Technical delivery conditions, *European Committee for Standardization*, Brussels, Belgium.

CEN (2005a). EN 1993-1-1:2005 – Eurocode 3: Design of steel structures - Part 1-1: General rules and rules for buildings, *European Committee for Standardization*, Brussels, Belgium.

CEN (2005b). EN 1993-1-2:2005 – Eurocode 3: Design of steel structures - Part 1-2: General rules – Structural fire design, *European Committee for Standardization*, Brussels, Belgium.

CEN (2005c). EN 1993-1-8:2005 – Eurocode 3: Design of steel structures - Part 1-8: Design of joints, *European Committee for Standardization*, Brussels, Belgium.

CEN (2005d). EN 1993-1-9:2005 – Eurocode 3: Design of steel structures - Part 1-9: Fatigue, *European Committee for Standardization*, Brussels, Belgium.

CEN (2005e). EN 1993-1-10:2005 – Eurocode 3: Design of steel structures - Part 1-10: Material toughness and through-thickness properties, *European Committee for Standardization*, Brussels, Belgium.

CEN (2006a). EN 10219-1:2006 – Cold formed welded hollow sections of non-alloy and fine grain steels - Part 1: Technical delivery conditions, *European Committee for Standardization*, Brussels, Belgium.

CEN (2006b). EN 10219-2:2006 – Cold formed welded hollow sections of non-alloy and fine grain steels - Part 2: Tolerances dimensions and sectional properties, *European Committee for Standardization*, Brussels, Belgium.

CEN (2006c). EN 10210-1:2006 – Hot finished structural hollow sections of non-alloy and fine grain steels – Part 1: Technical delivery conditions, *European Committee for Standardization*, Brussels, Belgium.

CEN (2006d). EN 10210-2:2006 – Hot finished structural hollow sections of non-alloy and fine grain steels – Part 2: Tolerances dimensions and sectional properties, *European Committee for Standardization*, Brussels, Belgium.

CEN (2006e). EN 1993-2:2006 – Eurocode 3 - Design of steel structures - Part 2: Steel Bridges), *European Committee for Standardization*, Brussels, Belgium.

CEN (2007a). EN 1993-1-12:2007 – Eurocode 3: Design of steel structures - Part 1-12: Additional rules for the extension of EN 1993 up to steel grades S 700, *European Committee for Standardization*, Brussels, Belgium.

CEN (2007b). EN 15048-1:2007 – Non-preloaded structural bolting assemblies - Part 1: General requirements, *European Committee for Standardization*, Brussels, Belgium.

CEN (2009a). EN 14399-9:2009 – High-strength structural bolting assemblies for preloading - Part 9: System HR or HV - Direct tension indicators for bolt and nut assemblies

CEN (2009b). EN 14399-10:2009 – High-strength structural bolting assemblies for preloading - Part 10: System HRC - Bolt and nut assemblies with calibrated preload, *European Committee for Standardization*, Brussels, Belgium.

CEN (2011). EN 1090-2:2008+A1:2011 – Execution of steel structures and aluminium structures - Part 2: Technical requirements for steel structures, *European Committee for Standardization*, Brussels, Belgium.

CEN (2013). EN ISO 898-1:2013 – Mechanical properties of fasteners made of carbon steel and alloy steel - Part 1: Bolts, screws and studs with specified property classes - Coarse thread and fine pitch thread (ISO 898-1:2013), *European Committee for Standardization*, Brussels, Belgium.

CEN (2014). EN ISO 5817:2014 – Welding - Fusion-welded joints in steel, nickel, titanium and their alloys (beam welding excluded) - Quality levels for imperfections (ISO 5817:2014), *European Committee for Standardization*, Brussels, Belgium.

CEN (2015). EN 14399-1:2015 – High-strength structural bolting assemblies for preloading - Part 1: General requirements, *European Committee for Standardization*, Brussels, Belgium.

Cerfontaine F (2003). *Study of the interaction between bending moment and axial force in bolted joints (in French)*, Ph.D. Thesis, Liège University.

CRIF, Université de Liege - Département M.S.M, CTICM Sait Rémy-les-Chevreuse, ENSAIS Strasbourg, Universita degli studi di Trento, LABEIN Bilbao. *Steel Moment Connections according to Eurocode 3, Simple design aids for rigid and semi-rigid joints*, Report SPRINT Contract RA 351.

Demonceau J-F (2008). *Steel and composite building frames: sway response under conventional loading and development of membrane effects in beams further to an exceptional action*, Ph.D. Thesis, Liège University.

Demonceau J-F, Hanus F, Jaspart J-P, Franssen JM (2009). *Behaviour of single-sided composite joints at room temperature and in case of fire after an earthquake*, International Journal of Steel Structures, Vol. 9(4), pp. 329-342.

Demonceau J-F, Haremza C, Jaspart J-P, Santiago A, Simões da Silva L (2013). *Composite joints under M-N at elevated temperatures*, In: *Proceedings of the Composite Construction VII Conference* (Composite Construction VII), Palm Cove, Australia.

Demonceau J-F, Hoang V-L, Jaspart J-P, Sommariva M, Maio D, Zilli G, Demofonti G, Ferino J, Pournara A-E, Varelis GE, Papatheocharis C, Perdikaris P, Pappa P, Karamanos SA, Ferrario F, Bursi OS, Zanon G (2012). *Performance-based approaches for high strength tubular columns and connections under earthquake and fire loadings (ATTEL)*, Report, RFCS Reserch Found for coal and steel, Brussels, http://bookshop.europa.eu/is-bin/INTERSHOP.enfinity/WFS/EU-Bookshop-Site/en_GB/-/EUR/ViewPublication-Start?PublicationKey=KINA25867.

Demonceau J-F, Jaspart J-P, Klinkhammer R, Weynand K, Labory F, Cajot LG (2008). *Recent developments on composite connections*, Steel Construction – Design and Research Journal, pp. 71-76.

Demonceau J-F, Jaspart J-P, Weynand K, Oerder R, Müller C (2011). *Connections with four bolts per horizontal row* (Eurosteel 2011), Budapest, Hungary.

329

DeWolf JT (1978). *Axially Loaded Column Base Plates*, Journal of the Structural Division ASCE, Vol. 104, pp. 781-794.

DeWolf JT, Ricker DT (1990). *Column Base Plates, Steel Design Guide* (Series 1), edited by AISC, Chicago.

Disque, Robert O (1984). *Engineering for Steel Construction*, American Institute of Steel Construction, Chicago.

Dubina D (2015). *High strength steel in seismic resistant building frames (HSS-SERF)*, (RFCS project - Final report), European Commission.

Dutch Commissie SG/TC-10a (Verbindingen) (1998). *Verbindingen: Aanbevelingen voor normaalkrachtverbindingen en dwarskrachtverbindingen*, Report not published.

ECCS (1994). *European Recommendations for Bolted Connections with Injection Bolts*, Publ. No. 79, ECCS - European Convention for Constructional Steelwork, Brussels, Belgium.

ESDEP. *Lecture 11.2.1.*

Ferrario F (2004). *Analysis and modelling of the seismic behaviour of high ductility steel-concrete compo-site structures*, Ph.D. Thesis, Trento University.

Franssen JM (2005). *SAFIR A thermal/structural program modelling structures under fire*, Engineering Journal AISC, pp. 143-158.

Gaboriau M (1995). *Recherche d'une méthode simple de prédimensionnement des ossatures contreventées à assemblages semi-rigides dans l'optique de l'approche élastique de dimensionnement (in French)*, Diploma work, University of Liège.

Gibbons C, Nethercot D, Kirby P, Wang Y (1993). *An appraisal of partially restrained column behaviour in non-sway steel frames*, Proc. Instn Civ. Engrs Structs & Bldgs, 99, pp. 15-28.

Gresnigt N, Romeijn A, Wald F, Steenhuis M (2008). *Column bases in shear and normal force*, HERON, Vol. 53, pp. 145-166.

Gresnigt N, Sedlacek G, Paschen M (2000). *Injection Bolts to Repair Old Bridges*, In: *Proceeding of the Fourth International Workshop on Connections in Steel Structures* (Connection in Steel Structures IV), edited by R Leon and W S Easterling, American Institute of Steel Construction (AISC), Roanoke, Virginia, USA, https://www.aisc.org/content.aspx?id=3626.

Gross LG (1990). *Experimental study of gusseted connections*, Engineering Journal AISC, pp. 89-97.

Guillaume M-L (2000). *Development of an European procedure for the design of simple joints (in French)*, Diploma work, University of Liège / CUST Clermont-Ferrand.

Hawkins MN (1968a). *The bearing strength of concrete loaded through flexible plates*, Magazine of Concrete Research, Vol. 20, pp. 95 -102.

Hawkins MN (1968b). *The bearing strength of concrete loaded through rigid plates*, Magazine of Concrete Research, Vol. 20, pp. 31-40.

Hoang V-L, Jaspart J-P, Demonceau J-F (2013). *Behaviour of bolted flange joints in tubular structures under monotonic, repeated and fatigue loadings I: Experimental tests*, Journal of Constructional Steel Research, Vol 85, pp. 1–11.

Hoang V-L, Jaspart J-P, Demonceau J-F (2014). *Hammer head beam solution for beam-to-column joints in seismic resistant building frames*, Journal of Constructional Steel Research, Vol. 103, pp. 49-60.

Jaspart J-P (1991). *Etude de la semi-rigidité des noeuds poutre-colonne et son influence sur la résistance et la stabilité des ossatures en acier*, Ph.D. Thesis, Department MSM, University of Liège.

Jaspart J-P (1997). *Recent advances in the field of steel joints. Column bases and further configurations for beam-to-column joints and beam splices*, Habilitation, Department MSM, University of Liège.

Jaspart J-P, Demonceau J-F (2008a). *Composite joints in robust building frames*, In: *Proceedings of the sixth International Composite Construction in Steel and Concrete Conference* (Composite Construction in Steel and Concrete VI), Colorado, USA.

Jaspart J-P, Demonceau J-F, Renkin S, Guillaume ML (2009). *European recommendations for the design of simple joints in steel structures*, edited by ECCS TC 10, Publ. No. 126, ECCS - European convention for constructional steelwork, Brussels, Belgium.

Jaspart J-P, Pietrpertosa C, Weynand K, Busse E, Klinkhammer R (2005). *Development of a Full Consistent Design Apprach for Bolted and Welded Joints - Application of the Component Method*, Report 5BP-4/05, CIDECT - Comité International pour le Développement et l'Etude de la Construction Tubulaire.

Jaspart J-P, Vandegans D (1998). *Application of the component method to column bases*, Journal of Constructional Steel Research, Vol. 48, pp. 89-106.

Jaspart J-P, Wald F, Weynand K, Gresnigt N (2008b). *Steel column base classification*, HERON, Vol. 53, pp. 127-144.

Jaspart J-P, Weynand K (2015). *Design of Hollow Section Joints using the Component Method, In: Proceedings of the 15th International Symposium on Tubular Structures, Rio de Janeiro, 27-29 May 2015* (ISTS 15), edited by E Batista, P Vellasco and L Lima, Rio de Janeiro, Brazil.

Jordão S (2008). *Behaviour of internal node beam-to-column welded joint with beams of unequal height and high strength steel*, Ph.D. Thesis, University of Coimbra.

Ju S-H, Fan C-Y, Wu GH (2004). *Three-dimensional finite elements of steel bolted connections*, Engineering Structures 26, pp. 403-413.

Kuhlmann U, Rölle L, Jaspart J-P, Demonceau J-F, Vassart O, Weynand K, Ziller C, Busse E, Lendering M, Zandonini R, Baldassino N (2009). *Robust structures by joint ductility*, Report EUR 23611 EN, RFCS Research, European Commission, Brussels, http://bookshop.europa.eu/is-bin/INTERSHOP.enfinity/WFS/EU-Bookshop-Site/en_GB/-/EUR/ViewPublication-Start?PublicationKey=KINA23611.

Kurobane Y, Packer JA, Wardenier J, Yeomans N (2004). *DG 9: Design guide for structural hollow section column connections, Construction with hollow sections*, edited by CIDECT - Comité International pour le Développement et l'Etude de la Construction Tubulaire, TÜV-Verlag GmbH, Köln, Germany.

Liew RJY, Teo TH, Shanmugam NE (2004). *Composite joints subject to reversal of loading - Part 2: analytical assessments*, Journal of Constructional Steel Research, pp. 247-268.

Maquoi R, Chabrolin B (1998). *Frame Design Including Joint Behaviour*, EU 18563 EN, Commision of The European Communities, Brussels, Belgium, http://bookshop.europa.eu/en/frame-design-including-joint-behaviour-pbCGNA18563/.

Moreno EN, Tarbé C, Brown D, Malik A, BCSA/SCI Connections Group (2011). *Joints in Steel Construction: Simple Joints to Eurocode 3*,

SCI Publ. P358, The Steel Construction Institute and The British Constructional Steelwork Association.

Murray TM (1983). *Design of Lightly Loaded Steel Column Base Plates*, Engineering Journal AISC, Vol. 20, pp. 143-152.

NEN 6770 (1997). *TGB 1990 - Staalconstructies - Basiseisen en basisrekenregels voor overwegend statisch belaste constructies.* Published by NEN (Nederlandse Normalisatie-instituut),

Nogueiro P, Simões da Silva L, Bento R, Simões R (2009). *Calibration of model parameters for the cyclic response of end plate beam-to-column Steel-concrete composite joints*, Steel and Composite Structures Journal, Vol. 9, pp. 39-58.

Owens GW, Cheal BD (1989). *Structural Steelwork Connections*, Butterworth & Co. Ltd.

Packer JA (1996). *Nailed tubular connections under axial loading*, Journal of Structural Engineering, Vol. 122, pp. 458-467.

Packer JA, Henderson JE (1997). *Hollow structural section connections and trusses: A design guide*, 2nd Ed., in, Canadian Institute of Steel Constructions, Willowdale, Ontario.

Packer JA, Wardenier J, Kurobane Y, Dutta D, Yeomans N (1992). *DG 3: Design guide for rectangular hollow section (RHS) joints under predominantly static loading, Construction with hollow sections*, edited by CIDECT - Comité International pour le Développement et l'Etude de la Construction Tubulaire, Verlag TÜV Rheinland GmbH, Köln, Germany.

Packer JA, Wardenier J, Zhao X-L, van der Vegte A, Kurobane Y (2009). *DG 3: Design guide for rectangular hollow section (RHS) joints under predominantly static loading*, 2nd Ed., in: *Construction with hollow sections*, edited by CIDECT - Comité International pour le Développement et l'Etude de la Construction Tubulaire, www.cidect.com.

Penserini P, Colson A (1989). *Ultimate Limit Strength of Column-Base Connection*, Journal of Constructional Steel Research, Vol. 14, pp. 301-320.

Renkin S (2003). *Development of an European process for the design of simple structural joint in steel frames (in French)*, Diploma work, University of Liège.

333

SCI (1995a). Joints in steel construction. Moment connections, P207, *The Steel Construction Institute.*

SCI (1995b). Advisory Desk Notes: Bending Moment in a pin, AD 172, *The Steel Construction Institute.*

SCI (2006). Column base stiffness for global analysis, NCCI-SN045a, *The Steel Construction Institute.*

SCI (2013b). Interim Report: Design of Portal Frames to Eurocode 3: An overview for UK designers, P400, *The Steel Construction Institute.*

Sedlacek G, Weynand K, Oerder S (2000). *Typisierte Anschlüsse im Stahlhochbau, Band 1: Gelenkige I-Trägeranschlüsse,* Band 1, Stahlbau Verlagsgesellschaft mbH, Düsseldorf, Germany.

Shelson W (1957). *Bearing Capacity of Concrete,* Journal of the American Concrete Institute, Vol. 29, pp. 405-414.

Simões da Silva L, Santiago A, Villa Real P (2011). *A component model for the behaviour of steel joints at elevated temperatures,* Journal of Constructional Steel Research, Vol. 57, pp. 1169-1195.

Simões da Silva L, Simões R, Gervásio H (2010). *Design of Steel Structures, ECCS Eurocode Design Manuals,* edited by ECCS - Europen Convention for Constructural Steelwork, Wilhelm Ernst & Sohn Verlag GmbH & Co KG, Berlin.

Stark JWB (1995). *Criteria for the use of preloaded bolts in structural joints,* In: *Proceeding of 3rd International Workshop on Connections* (Connection in Steel Structures III), edited by R Bjorhovde, A Colson and R Zandonini, Pergamon, Trento, Italy.

Steenhuis CM (1998). *Assembly Procedure for Base Plates,* Report 98-R-0477, TNO Building and Construction, Delft

Steenhuis M, Bijlaard FSK (1999). *Tests On Column Bases in Compression,* In: *Commemorative Publication for Prof. Dr. F. Tschemmemegg* (Timber and Mixed Building Technology), edited by G Huber.

Steenhuis M, Wald F, Sokol Z, Stark JWB (2008). *Concrete in compression and base plate in bending,* HERON, Vol. 53, pp. 109-126.

Stockwell FW, Jr (1975). *Preliminary Base Plate Selection,* Engineering Journal AISC, Vol. 21, pp. 92-99.

Togo T (1967). *Experimental study on mechanical behaviour of tubular joints*, Ph.D. Thesis, Osaka University.

Veljkovic M (2015). *Eurocodes Workshop: Design of Steel Buildings with Worked Examples*.

Wald F (1993). *Column-Base Connections, A Comprehensive State of the Art Review*, Report, CTU, Praha.

Wald F, Obata M, Matsuura S, Goto Y (1993). *Flexible Baseplate Behaviour using FE Strip Analysis*, Acta Polytechnic a CTU, Prague, Vol. 33, pp. 83-98.

Wald F, Sokol Z, Jaspart J-P (2008a). *Base plate in bending and anchor bolts in tension*, HERON, Vol. 53, pp. 79-108.

Wald F, Sokol Z, Steenhuis CM, Jaspart J-P (2008b). *Component method for steel column bases*, HERON, Vol. 53, pp. 61-78.

Wardenier J (1982). *Hollow section joints*, Delft University Press, Delft, The Netherlands.

Wardenier J (2001). *Hollow sections in structural applications*, edited by CIDECT - Comité International pour le Développement et l'Etude de la Construction Tubulaire, www.cidect.com.

Wardenier J, Kurobane Y, Packer JA, Dutta D, Yeomans N (1991). *DG 1: Design guide for circular hollow section (CHS) joints under predominantly static loading, Construction with hollow sections*, edited by CIDECT - Comité International pour le Développement et l'Etude de la Construction Tubulaire, Verlag TÜV Rheinland, Köln, Germany.

Wardenier J, Kurobane Y, Packer JA, van der Vegte A, Zhao X-L (2008). *DG 1: Design guide for circular hollow section (CHS) joints under predominantly static loading*, 2nd Ed., in: *Construction with hollow sections*, edited by CIDECT - Comité International pour le Développement et l'Etude de la Construction Tubulaire, www.cidect.com.

Wardenier J, Packer JA, Zhao X-L, van der Vegte A (2010). *Hollow sections in structural applications*, edited by CIDECT - Comité International pour le Développement et l'Etude de la Construction Tubulaire, Bouwen met Staal, Zoetermeer, The Netherlands.

Weynand K, Jaspart J-P, Demonceau J-F, Zhang L (2015). *Component method for tubular joints*, Report 16F - 3/15, CIDECT - Comité International pour le Développement et l'Etude de la Construction Tubulaire.

Weynand K, Kuck J, Oerder R, Herion S, Fleischer O, Rode M (2011). *Design Tools for Hollow Section Joints*, 2nd Ed., in, V&M Deutschland GmbH, Düsseldorf, Germany, www.vmtubes.com.

Weynand K, Oerder R (2013). *Standardised Joints in Steel Structures to DIN EN 1993-1-8*, Complete Edition 2013 Ed., in, Stahlbau Verlags- und Service GmbH, Düsseldorf, Germany.

Weynand K, Oerder R, Demonceau J-F (2012). *Resistance tables for standardized joints in accordance with EN 1993-1-8*, In: *Proceeding of 7th International Workshop on Connections* (Connection in Steel Structures VII), Timosoara, Rumania.

Weynand K, Oerder R, Jaspart J-P (2014). *COP - The Connection Program*, Program to design steel and composite joints in accordance with EC3 and EC4, Version 1.7.2, Feldmann + Weynand GmbH, Germany, http://cop.fw-ing.com.

Whitmore RE (1952). *Experimental investigation of stresses in gusset plates*, Report Bulletin 16, Engineering Experiment Station, University of Tennessee.

Zoetemeijer P (1974). *A design method for the tension side of statically loaded bolted beam-to-column connections*, Vol. 20, Heron, Delft, The Netherlands.

Annex A

Practical Values for Required Rotation Capacity

The following Table gives some practical values for required rotation capacity $\phi_{required}$.

System of loading	M_{max}	$\phi_{required}$
  A B M	M	$\theta_A = \dfrac{\gamma ML}{6EI}$ $\theta_B = -\dfrac{\gamma ML}{3EI}$
  P	$\dfrac{PL}{4}$	$\pm\dfrac{\gamma PL^2}{16EI}$
  P	$\dfrac{pL^2}{8}$	$\pm\dfrac{\gamma pL^3}{24EI}$
  Total loading P L	$\dfrac{2PL}{9\sqrt{3}}$	$\theta_A = \dfrac{7\gamma PL^2}{180EI}$ $\theta_B = \dfrac{8\gamma PL^2}{180EI}$

Where

    E is the elastic modulus of the material from which the beam is formed;

    I is the second moment area of a beam;

    L is the span of a beam (centre-to-centre of columns);

    γ is the loading factor at ULS.

Annex B

Values for Lateral Torsional Buckling Strength of a Fin Plate

Lateral torsional buckling strength of a fin plate f_{pLT} (BS 5950-1 Table 17 (BSI, 2000)).

λ_{LT}	Steel grade and design strength f_y (N/mm^2)														
	S275					S355					S460				
	235	245	255	265	275	315	325	335	345	355	400	410	430	440	460
25	235	245	255	265	275	315	325	335	345	355	400	410	430	440	460
30	235	245	255	265	275	315	325	335	345	355	390	397	412	419	434
35	235	245	255	265	272	300	307	314	321	328	358	365	378	385	398
40	224	231	237	244	250	276	282	288	295	301	328	334	346	352	364
45	206	212	218	224	230	253	259	265	270	276	300	306	316	321	332
50	190	196	201	207	212	233	238	243	248	253	275	279	288	293	302
55	175	180	185	190	195	214	219	223	227	232	251	255	263	269	281
60	162	167	171	176	180	197	201	205	209	212	237	242	253	258	269
65	150	154	158	162	166	183	188	194	199	204	227	232	242	247	256
70	139	142	146	150	155	177	182	187	192	196	217	222	230	234	242
75	130	135	140	145	151	170	175	179	184	188	207	210	218	221	228
80	126	131	136	141	146	163	168	172	176	179	196	199	205	208	214
85	122	127	131	136	140	156	160	164	167	171	185	187	190	192	195
90	118	123	127	131	135	149	152	156	159	162	170	172	175	176	179
95	114	118	122	125	129	142	144	146	148	150	157	158	161	162	164
100	110	113	117	120	123	132	134	136	137	139	145	146	148	149	151
105	106	109	112	115	117	123	125	126	128	129	134	135	137	138	140
110	101	104	106	107	109	115	116	117	119	120	124	125	127	128	129
115	96	97	99	101	102	107	108	109	110	111	115	116	118	118	120
120	90	91	93	94	96	100	101	102	103	104	107	108	109	110	111

B. VALUES FOR LATERAL TORSIONAL BUCKLING STRENGTH OF A FIN PLATE

λ_{LT}	Steel grade and design strength f_y (N/mm^2)														
	S275					S355					S460				
	235	245	255	265	275	315	325	335	345	355	400	410	430	440	460
125	85	86	87	89	90	94	95	96	96	97	100	101	102	103	104
130	80	81	82	83	84	88	89	90	90	91	94	94	95	96	97
135	75	76	77	78	79	83	83	84	85	85	88	88	89	90	90
140	71	72	73	74	75	78	78	79	80	80	82	83	84	84	85
145	67	68	69	70	71	73	74	74	75	75	77	78	79	79	80
150	64	64	65	66	67	69	70	70	71	71	73	73	74	74	75
155	60	61	62	62	63	65	66	66	67	67	69	69	70	70	71
160	57	58	59	59	60	62	62	63	63	63	65	65	66	66	67
165	54	55	56	56	57	59	59	59	60	60	61	62	62	62	63
170	52	52	53	53	54	56	56	56	57	57	58	58	59	59	60
175	49	50	50	51	51	53	53	53	54	54	55	55	56	56	56
180	47	47	48	48	49	50	51	51	51	51	52	53	53	53	54
185	45	45	46	46	46	48	48	48	49	49	50	50	50	51	51
190	43	43	44	44	44	46	46	46	46	47	48	48	48	48	48
195	41	41	42	42	42	43	44	44	44	44	45	45	46	46	46
200	39	39	40	40	40	42	42	42	42	42	43	43	44	44	44
210	36	36	37	37	37	38	38	38	39	39	39	40	40	40	40
220	33	33	34	34	34	35	35	35	34	36	36	36	37	37	37
230	31	31	31	31	31	32	32	33	33	33	33	33	34	34	34
240	28	29	29	29	29	30	30	30	30	30	31	31	31	31	31
250	26	27	27	27	27	28	28	28	28	28	29	29	29	29	29
0	37.1	36.3	35.6	35.0	34.3	32.1	31.6	31.1	30.6	30.2	28.4	28.1	27.4	27.1	26.5

where $\lambda_{LT} = 2.8\left(\dfrac{z_p h_p}{1.5 t_p^2}\right)^{1/2}$